Ethical
Hacking

Learn About Effective Strategies of Ethical Hacking

(A Complete Beginners Guide to Successful Ethical Hacking Career)

John Ellis

Published By **Ryan Princeton**

John Ellis

All Rights Reserved

Ethical Hacking: Learn About Effective Strategies of Ethical Hacking (A Complete Beginners Guide to Successful Ethical Hacking Career)

ISBN 978-1-77485-787-8

No part of this guidebook shall be reproduced in any form without permission in writing from the publisher except in the case of brief quotations embodied in critical articles or reviews.

Legal & Disclaimer

The information contained in this ebook is not designed to replace or take the place of any form of medicine or professional medical advice. The information in this ebook has been provided for educational & entertainment purposes only.

The information contained in this book has been compiled from sources deemed reliable, and it is accurate to the best of the Author's knowledge; however, the Author cannot guarantee its accuracy and validity and cannot be held liable for any errors or omissions. Changes are periodically made to this book. You must consult your doctor or get professional medical advice before using any of the suggested remedies, techniques, or information in this book.

Upon using the information contained in this book, you agree to hold harmless the Author from and against any damages, costs, and expenses, including any legal fees potentially resulting from the application of any of the

TABLE OF CONTENTS

Introduction

If you are looking to hack another person's system to obtain information illegally, please stop reading this book right away. You should continue to read the book if you want to learn more about how to test the vulnerabilities in a system or network and want to fix those vulnerabilities. This book provides information on different techniques an ethical hacker can use to identify any vulnerabilities in a system or network and identify a way to fix those vulnerabilities. Most organizations perform this exercise to prevent a malicious hack on the organization's network and infrastructure. This book only talks about ethical hacking, which is a legal way of testing the vulnerabilities in a system. You must understand that both computer and network security constantly evolve. Therefore, you must ensure that you always secure your computer and network from criminal hackers or crackers.

This book lists different tools and techniques that you can use to test the system or network for any vulnerabilities. Once you identify the vulnerabilities, you can work towards improving network security. If you do not know how a hacker thinks, you may not be able to test the system well. If this is the case for you, then you should spend some time to understand how a hacker thinks and use that knowledge when you are assessing the system.

Ethical hacking is also called penetration testing or white hat hacking, and it is used by many organizations to ensure that their network and systems are secure. This book will provide information about different software and tools that you can use when you are performing an ethical hack. There are some sample exercises and programs in the book that you can use to begin the ethical hacking process.

I hope you are able to gather all the information you need from this book. Once again, please refrain from using the content within for any illegal purposes.

Chapter 1: Fighting Against Companies

Imagine two giants fighting out in the open. They could be swinging trees at one another or hurling entire hills like pebbles. Anyone caught in the crossfire will have a bad day, but nearby towns could also get trampled in a second, and the entire landscape is sure to be completely destroyed as a consequence of this fight. The giants don't care about the little people scurrying around, unless they start posing a threat or draw too much attention to themselves. Giants would also not obey kings, councils, moral standards or any rules of fair play, though they might pretend to do so when it suits them. After all, would you follow rules set by people 50,000 times smaller than you?

We live in a world dominated by companies that are just like these giants. They may seem intimidating until you realize that they're also slow, stupid, half-blind and utterly predictable. In fact, the only thing these giants have going for them is their overwhelming strength, which they can't even use properly, randomly causing chaos and destruction wherever they step. Thinking their power can be harnessed is utter lunacy and yet governments all around the world keep trying to do so, offering deals that ultimately hurt the little people. By working with the government, tech giants can get unprecedented privileges, both legal and monetary, while the government can exercise influence without constitutional boundaries. Regular users are just an afterthought, a statistic that's useful in the aggregate as a bragging right but are otherwise left to themselves.

Throughout this book, you'll be presented with arguments showing how all companies eventually become giants, trampling the people they're supposed to be working for. You'll also learn how the only recourse is for each of us to independently find ways to live a relatively sheltered life and slightly weaken these companies until they decide to play fair. No organization or legislation can truly rein the giants in, but a group of determined people who are small, independent and practically invisible can, and in doing so also make their lives just a tiny bit more comfortable. This is where you come in as you vote with your wallet and use your brain for the benefit of all of humanity.

Ethical hacking is perhaps the most thrilling way to use any given product, but we have to keep our wits about us and not abuse what we know to become villains. Techniques will be described theoretically as much as possible to avoid censorship as hacking is still a taboo topic in the mainstream thanks to popular culture. If you want to work as an ethical hacker, keep in mind that the way you bring value to companies is by finding security loopholes before someone with malicious intentions find them.

There are skill levels to ethical hacking, ranging from simple social engineering to elaborate programming. This book won't presume you prefer one or the other or have an aptitude in any of them. A genius programmer might be a stuttering mess and a social butterfly might type with one finger and yet they both have the same chances of being successful hackers. Whatever your skills are, you should work on them and refine them during your entire life because there's money to be made hacking.

Computers are increasingly involved in our lives, so knowing how and why they work can prevent and solve a great many problems you would otherwise need to pay someone else to fix. Any hacking skill you learn over the course of your life can be used to make money. It's just that with ethical hacking you can do it legally, by improving products, customizing services and using those skills to teach and tutor others in person and over the internet. Like all decent teaching materials, this book will give you a glimpse into what's possible to whet your learning appetite – the rest is up to you.

Chapter 2: Ethical Hacking Defined

How do you imagine hackers? Do they wear sunglasses and hoodies indoors and type on the keyboard faster than the eye can see? Do they drink Jolt Cola while trying to "hack the Gibson"? In all likelihood, you'd probably never recognize an actual hacker passing you on the street. They could be just some pimply girl who hacked into a multimillion-dollar company on a dare. As this book is about to show you, there's nothing easier than becoming a hacker, and you can be one as well. The trick is in becoming an ethical hacker, a special breed of hacker that does what he can to extract maximum utility out of products and services but doesn't abuse or exploit others.

In broadest terms, hacking is unauthorized modification of a product to allow unintended use. For example, a car might have its turn signal blinkers hacked to allow them to transmit messages in Morse code. Passive voice isn't there merely to make English majors froth at the mouth; truly anyone can hack a car. This would be made possible because the car manufacturer created an elaborate software package and

7

baked it into the car, which also happens to have plenty of holes and ways to hack it.

Adding unnecessary functionality to products increases their attack surface, the likelihood of getting hacked, for marginal benefit. Companies feel compelled to do it because it's the latest trend, or their competition is also doing it and they lack faith in their own products. None of this is supposed to happen; engineers making the products certainly didn't envision you pulling your hair out trying to use their printer. The thing is, engineers that make products are separated from the public by layers of marketers and businessmen who have their own agenda, namely, to hype the product up and extract maximum revenue out of customers.

So, the blinkers can transmit Morse code but it's a dormant capability that the car maker didn't consider or couldn't disable without disabling the blinkers themselves. That's the key concept when it comes to hacking—it's so simple, cheap and straightforward that it can't be stopped. Another valuable concept is having physical ownership of the hardware; if you have unfettered access to the car, computer or software in question, defeating whatever security features it has is only a matter of time.

Hacking tools do make it easier to perform sophisticated attacks, in particular when the hacker wants to remain undetected, but a bare-bones hack can be done using just – a plain sticker. For example, a state-of-the-art self-driving car using a neural network can be confused by slapping stickers on traffic signs[1], which doesn't impede human performance but makes the car software go haywire. Here we encounter the golden rule of all hacking, which is code is law, or "whatever hackers can do with any given product determines its ultimate function." Companies respond to this by using actual law to sue everyone interfering with their products in the open.

But why are products so vulnerable to hacking? How come car software isn't protected with hundreds of layers of security, including barbed wire and booby traps? First of all, companies making cars want to push out a new product as soon as possible, which means putting the least amount of effort into it. Second, adding new systems means more money spent on car design, which cuts into the company's profit margin. Third, in all likelihood,

customers won't even know or care how the car runs as long as it does, making all but the most basic security pointless. We can term this rationale "the dumb customer rule" and describe it as "companies want to get rid of their customers until only the dumbest remain". Fourth, all security measures are temporary and get defeated anyway in due time, so security is just delaying the inevitable. Finally, adding too much security might lead to customers experiencing injury or damage, opening the company to lawsuits and negative publicity.

With the advent of cloud-connected self-driving cars, we might witness total mayhem as Chinese, Russian or Zimbabwean hackers wreak havoc on streets and parking lots half a globe away. As cyberspace and physical world merge, we will start getting internet access on plenty of things that never had them before, opening us all to hacking attacks. Becoming an ethical hacker thus becomes a necessity born out of sheer frustration, a way to infuse some stability into our lives on our own and fight off pesky cyber-attackers.

Hackers could be fiddling with products out of curiosity, which would make them white hat, or out of malice, making them black hat. Curiosity is a peculiar urge we don't consider all that much, but some white hats truly are driven to pick things apart and watch them tick with childlike joy. On the other hand, black hats can be causing mayhem to make money, steal data or just revel in the chaos they caused. In this book, we'll discuss the third kind of hacking, the idea of ethical hacking, a utility-driven effort to make products and services work as advertised, or as expected. Looking at hacking from a utility standpoint, black hat would want to diminish utility, white hat would want to not influence it at all and an ethical hacker would want to increase utility.

The sad thing is that white hats generally make no efforts to hide their hacking while black hats do. When law enforcement finally gets to unload their wrath on someone for all the hacking, it's usually a hapless white hat. Legally, both types of hacking are considered exactly the same, contrary to all established legal practice that considers intention as a key element for a crime, which is incidentally why animals can't ever be charged with murder. Ethical hacking therefore only makes sense if you're not exposing yourself to risk, which should be your first priority. Just don't do anything that could get you into legal trouble and don't brag about your exploits over the internet, and you should be fine.

Chapter 3: War On The Internet

Speaking of which, the internet itself is a US military invention created sometime around 1970 as ARPANET, a hardened communication line between underground command centers in case of a nuclear war with the USSR. With the ability to route around damaged parts, ARPANET used a system of trusted relays to instantly get messages and files across. That's how the internet works today, and there's plenty of evidence that the civilian infrastructure was simply built on top of ARPANET. The whole shebang is still being used as an economic, information and political battleground, with civilians blissfully unaware of what's going on.

Your devices connected to the internet could be hacked into by black hats to be used against you or recruited to wage a cyber-war, and you'd be none the wiser. Disconnect a device from the internet when you're not using it and never reuse or reveal your passwords to anyone in person or online. In general, longer passwords made out of simpler words are much stronger than shorter but convoluted passwords.

Malware also seems to be used as a weapon, and cases like Stuxnet, Duqu and Flame are arguably the products of US cyber-warfare efforts, in particular against Iran[2]. Black hats eventually get ahold of this kind of malware and use it for stealing data, identities or just for causing mayhem—like in real warfare, civilians are just collateral damage. There's no bulletproof way to stop malware so back up your data on separate hardware in case the worst happens.

Tech giants are involved in waging war as well, with Google employees staging a protest in April 2018 to persuade their CEOs to back away from a Pentagon deal[3] that would have them work on improving autonomous drones' targeting capabilities. Even without weapons, tech giants that manage online platforms hold the scary power of being able to completely de-platform individuals for their political views or for no reason at all. We need not agree with their views, but the very notion of that happening to anyone should be heinous because we could be next. What's worse is that companies inevitably start leaning left once they become big enough, engaging in thought control to ensure uniformity and pushing that view on their users. The first victims of this culture war are always capable employees.

The case of James Damore[4], a brilliant Google engineer who was fired because he wrote a memo about men and women differing in their capabilities, shows how tech giants have a knee-jerk reaction to anything that might be seen as against ultra-liberal views of equality. Offering an alternative platform in such circumstances becomes impossible; the mainstream media, tech giants and even legislative bodies can join forces to maintain status quo.

Gab[5] is an attempt at providing an alternative to Twitter that shows what happens when people decide to spite tech giants. Gab is still online but attracted so much vitriol from Silicon Valley companies that its days can be considered numbered. Based in Texas and having no ads, Gab touts itself as a free speech platform, unlike Twitter that uses automation and moderators to keep things clean according to its standards.

After a suspect in a synagogue shooting was found to have had posts on Gab, PayPal suspended its account[6], effectively cutting Gab's jugular. Google and Apple banned Gab's app from their stores, citing pornography and hate speech. Microsoft also threatened banning Gab from their Azure hosting service due to "threats of ritual murder". Controversial material exists on other social networks too, but it's only when on Gab that it's used as an excuse to attack the platform. Anyone trying to create a similar alternative can expect to be slandered and beset from all sides in a similar manner.

This suggests a scary future in which tech giants coordinate to create a list of non-persons, people who have said the wrong thing and are thus comprehensively banned from all their products and services. If you're banned by Google, Twitter, PayPal and Microsoft, do you even exist? What can you do about it? If the media joins in on the side of tech giants, as history has shown they tend to do, there won't be a way to get your side of the story out. However, you can avoid the problem in the first place if you think small.

Dominant social networks and platforms are too big for their own sake, exposing their users to privacy intrusions and hacking attacks. Arguably, they don't work as advertised either, since feature creep, tendency of software to extend way beyond its original purpose, changes the way we use it. Any alternative to Facebook, Twitter or Instagram that truly wants to be dedicated to free speech will be decentralized, scalable, cheap, host up to 100 users and, most importantly, be impossible to censor. The only problem is that its user interface will be bland though functional, repulsing those addicted to slick animations and colorful icons. If tech giants decide to stamp it out, the same platform will be replicable with the tiniest of investments.

This also implies that there won't be a recognizable brand to slander in the mainstream media, no keyword to filter out from e-mails, messages and search results and no way to block the flow of micro-payments and data to and from it. A microbe amongst giants, such a social network would be invisible and unstoppable. Each user would be able to customize his own version and deploy it almost instantly for others to find. In other words, it would be truly anonymous and there wouldn't be profiles to be made unless each user specifically sets out to make himself memorable. This isn't advised since you never know who's looking.

Consider everything you do online to be tracked, cataloged and analyzed, not by people, but by machines with artificial intelligence perfected using machine learning, a process of digital evolution. They will be the perfect arbiters of decency at some point in the future, or so the tech giants would like us to think. Just like humans can err, these digital brains can also experience fatigue and small missteps can accumulate to produce disasters further down the line.

Working a million times faster than humans, there won't be anyone capable of pulling the plug before they bring down the biggest websites that employ them with a spectacular crash. Amidst all of this, you just want to have a working product, some semblance of peace, quiet and utility. This attitude is the most sensible and forward-thinking one you can imagine, and all it takes is for you to start thinking like an engineer.

So, who's going to make this free speech competitor to tech giants and their mind control agenda? Why not you? Learning how to code can be started with nothing more than Notepad or any other word processing program. There are plenty of tutorials online, though you might want to start off with something more accessible, such as AutoHotkey[7], a powerful scripting program that normally produces special text files the program can not only parse but can produce stand-alone executables as well.

If you do any subversive work on creating a free speech platform, do cover your tracks as Satoshi Nakamoto did. The fabled creator of Bitcoin, nobody knows if Satoshi was a man, woman, AI or intelligent crystal because he did not expose any personal information. He set the foundation for what's now a booming cryptocurrency market, mined about a million Bitcoin for himself and disappeared without a trace. If he were a public representative of Bitcoin, plenty of people would try to control or harm him, but all we have are a couple dozen forum posts.

Chapter 4: Engineer's Mind

Engineers are a funny breed in that they have a fairly cold, calculating and clinical approach to solving problems. For an engineer, presentation doesn't matter; an ugly solution is still a solution and might even be the optimal one because no time was spent puffing it. When employed by a company to produce something like software, a tractor or a smartphone, an engineer will look at material constraints, budget, deadline and project parameters to find whatever fits all of those. The implication of this approach is that the product made by an engineer will necessarily have severe flaws because it's impossible to account for every possible event unless the budget is infinite, which it never is.

In the case of the Ford Pinto, the problem was that the fuel tank was situated right behind the rear bumper with an axle and its studs between the two. Any impact to the rear bumper could rupture the gas tank and cause a fire. A high-speed collision would also deform the front doors as a result of the force, making escape impossible. When three teenage girls riding in a Ford Pinto died in a fiery rear-end collision in 1978, the Ford company was charged with murder and recalled 1.5 million vehicles[8], even though their safety was statistically no better or worse than for other cars on the market. Compare this to Volvo, which has made nearly indestructible cars[9] that tear through other vehicles like wet cardboard but isn't getting nearly the kind of recognition its safety deserves.

The general public has an emotionally charged way of looking at things, as they showed by blaming Ford for the death of three young girls. If you adopt that same attitude, you will get hurt without accomplishing anything and even thinking like an engineer might not yield results. James Damore wrote his memo from a standpoint of an engineer and got fired regardless. Gab founders thought like engineers in making an alternative to Twitter and still got besmirched. You can't win by throwing a temper tantrum or by simply butting heads on an engineering level, but you can employ some sneaky ethical hacking, which could be as simple as ditching Chrome and adopting another browser that is more suitable to your newfound needs. That's it--you're not changing the world, but your world.

So, the company employing the engineer looks at the same product from both a marketing and business standpoint. Does the product make the world a better place and does it make the company any money. Bad publicity hurts the company just as much as the product not selling at all, but it's nearly impossible to just generate goodwill. So companies generally go for the most risk-averse strategy that won't cause lawsuits and bad press, regardless of if the product is actually defective.

An ideal customer for the company would be one that's too dependent to ever leave and just rich enough to be a worthwhile target but not rich enough to fight back legally. When the company needs a cash injection, terms of service can be changed to increase fees, even if it's by 1%. Usually the company will give advance notice of 30 or 60 days and blithely state that those who don't agree may leave. However, the grace period doesn't matter if your business depends on that service; you'll pay up because that's always preferable to shuttering your doors. As long as the company thinks the customers are dumb enough to not realize their options, it will always be tempted to randomly extort more money for the same value. Thus, all companies gravitate towards the golden business principle—don't make a better product, find dumber customers.

The qualifier "dumb" isn't meant to make anyone feel bad but simply to point out that companies loathe technically-capable users. If you can fix your own car, toaster or refrigerator, you've just denied companies that sell those products and might have authorized repair shops repeat business. You're also dangerously close to becoming competition to the company's own retailers or repair shops if you have any of those skills or just have the curiosity to find out how these products work. With competence comes understanding, in particular as to how much these products and their repairs are worth, giving you the ability to undercut and outmaneuver these companies simply by having the bare minimum of competence.

Customization is another important benefit of being a smart customer. When you can delve into the intricacies of any given product, you can also adapt it to your needs and circumstances, perhaps by making a web browser into a multi-functional tool or by hacking a toaster to store MP3 files. Those who can't customize a product have to deal with frustration as it inevitably fails to perform as expected. Even then, the company profits as frustration is a powerful emotion that can be exploited, such as by offering a free phone support line, keeping the customers waiting but also allowing the option to immediately get in touch with expert tech support for a small charge. Once you realize that companies benefit from causing frustration in their customers, you'll start seeing this principle in action everywhere around you.

An ethical hacker would thus want to explore the entire world and all the gadgets it's housing with a fresh set of eyes--ones that can see utility. The world is in constant need of elegant, simple and robust solutions for all sorts of problems, since plenty of our routines are based on experiences that are no longer valid but are still held dear. Just like every living thing is in a constant motion of self-perfection, ethical hackers can carefully guide mechanical things to a state of perfection or as close to it as possible. Of course, for companies, none of this makes sense. They just want to get money and not get sued or derided as socially irresponsible, meaning they want to lock their products down, preferably using a method that gives them total control over when, how and why their product is used.

Now you've started to see the intricacies of bringing any product to market. Companies also grow, change their goals, and over time, morph into something different. In general, this process is irreversible. A company will slowly attract managerial and accountant types that may be good with numbers but have no clue what made the company great. Lacking imagination or an adventurous spirit, these managers and accountants will slowly bleed the company dry and, if there's enough of them, make it follow market trends rather than set them. Meanwhile, the official narrative is that everything's swell, and the company is making the world a better place.

Each product you've ever bought is a result of these three parties interacting and vying for dominance. Engineers just want to make solid products, marketing just wants to make them glitzy and business people just want to make money. Depending on the imbalance between these forces in the company, the product can be useful but ugly and make the company broke, useless and pretty but make a ton of revenue or anything in between. Because marketing and management tend to wield more clout than engineers, the former two groups tend to control the company direction at the expense of product quality. There are exceptions to this, most notably Bill Gates' Microsoft and Steve Jobs' Apple.

Both Bill and Steve were notorious for having temper tantrums when asked to evaluate a half-baked product made by their engineering departments. In one case, Steve took an iPod and threw it into an aquarium to prove that it can be made smaller; rising air bubbles showed that it indeed can[10]. Bill would memorize license plates[11] of his employees' cars to know who was leaving work and when. In his own words, "I was quite fanatical about work." Both of them emphasized the engineering aspect of any given product, resulting in fantastic companies that have remained afloat all these years.

Mere words don't do justice to just how hectic this environment is. Fickle billionaires, backstabbing executives and outrageous parties are all a part of daily life when international corporations burn through millions of dollars in a struggle to stay relevant in the global market. If you've been watching Silicon Valley, there's a great moment where protagonists visit TechCrunch Disrupt[12], an actual pitch conference for startups. Buzzwords galore as startup owners promise to make the world a better place through things such as "Paxos Algorithms for consensus protocols" and "software-defined data centers for cloud computing", with the ultimate punchline to the scene being a fistfight worthy of Borat. The line between reality and parody blurs and the only flaw of the series is that Mike Judge, the creator of such masterpieces as Idiocracy, had to tone it down to make everything more believable.

Chapter 5: The Almighty Eula

EULA, or End User License Agreement, is that block of text we all ignore and click "I Agree" when installing a program on our desktop machine. A team of lawyers must have invested thousands of hours pruning and perfecting that same EULA down to the tiniest comma, and yet users simply skip over it, tut-tut. Just kidding, it was probably copied and pasted from a random website by an intern. Even Google was caught doing it, since one tidbit in Chrome EULA allowed them to publicly display any information transmitted through Chrome, which is fine as long as you don't use it for banking (hint: it was probably copied from a video game EULA, which had to announce public display of private messages for communication in a multiplayer game).

The thing is, a EULA is meant to govern all disputes arising from that software's use, but it's written by the company and will naturally tilt the terms in its favor. For example, any disputes are usually meant to be resolved in a Californian court, which is where Silicon Valley houses the majority of software makers in the US. Have fun flying over there to air your grievances over some software. EULAs are also written in all caps and displayed in a tiny window to make it even less likely someone will go through the trouble of reading it all the way.

By 2006, Microsoft's EULA for Windows XP got so strict that it allowed only one installation per user on a single machine, regardless of how many machines he had. Technically speaking, that EULA defined a machine as having a single processor, which could be interpreted to mean a user running a PC with multi-core architecture would have to buy a copy for each core. OEM (original equipment manufacturer) versions of Windows were sold too, and they allowed only one owner and machine per copy — ever.

Over time EULAs became so outrageous that one company even put in a $1,000 reward for anyone who contacted them at the email address buried underneath the legalese. Apple blundered with a EULA too when it released Safari for Windows and stated that it must be installed on an Apple machine, meaning the user had to run

Windows on an Apple. Common terms in EULAs prohibit making backups, which means the automated backup systems commonly found in Windows violate it; allow the issuing company to change it at any time and without prior warning, making whatever is already written in it meaningless; and forbid class-action lawsuits, which is likely impossible to agree to[13]. Far Cry 2, a sandbox video game, had a clause in its EULA that prohibited using the game "contrary to morality", whatever that means.

It is unclear whether EULA, aka shrink-wrap agreement, aka click-wrap agreement, is even legally binding in a court of law. The name stems from the idea that the EULA is considered binding as soon as the buyer opens the shrink-wrap packaging or clicks through the installation menu. The problem is that a typical contract usually involves a signed agreement where all parties had at least some chance to tweak the clauses. With EULA, it's doubtful clicking a button has the same weight as a signature and the customer has no way of seeing it before the purchase, effectively forcing acceptance.

EULA was proven rickety in court in a 1996 case of a student buying a phonebook in CD format from a company and putting up a website where he charged for access to the phone numbers. Since the company made money selling those CDs, they sued the guy.

Wisconsin district court ruled[14] that: a) copyright was not breached because phone numbers are already public information and b) the student did use the program for non-commercial purposes, as stated in the EULA. However, phone numbers didn't count; again, the company tried to set its terms to the use of public information. Note this detail, as we'll keep seeing this exact same pattern of what common property is being claimed by companies for their profit. The company appealed, and the US Court of Appeals overruled[15] the decision, disagreeing on both counts and stating that click-wrap contracts are always binding unless their terms would be wrong in regular contracts or if they are "unconscionable".

Microsoft had its share of lawsuits, primarily dealing with unauthorized distribution. In

1994, Microsoft sued[16] Harmony Computers, a small company that installed their MS-DOS operating system on computers and sold them. Harmony Computers argued that they bought their copies legitimately and were thus protected under the first sale doctrine that protects resellers directly supplied by the author. Microsoft responded that the first sale doctrine didn't apply because MS-DOS was licensed, not sold. Even if it did apply, Harmony Computers had to have bought their copies directly from Microsoft, which they didn't.

Go look through any EULA of any program or service, and you'll see that you're never a buyer, but a licensee, which gives that company excessive rights to meddle in your affairs. As a customer of these tech giants, you don't have rights, only privileges and those can be taken away as soon as you're deemed too stupid to fight back.

In 1998, the US government sued Microsoft for breaching anti-trust regulations and stifling free market competition by bundling Internet Explorer with Windows and DOS, among other things. During deposition, Bill Gates reportedly acted like a petulant child, refusing to answer questions and asking for definition of words such as "we". Bill constantly denied he remembered anything when the prosecutor would bring up snippets of his e-mails proving otherwise. To accusations that they were acting like a monopoly, Microsoft CEOs responded that the competition was simply jealous of their success. The sitting judge in that case ordered Microsoft broken up in two parts: one would make Windows and the other everything else. This ruling was appealed and overturned because Microsoft was already too big to fail, and its fragmentation would harm the entire US economy. How does one even begin to rein in a company as powerful as Microsoft?

In 2001, Microsoft entered a settlement with the US government, agreeing to have their software building tools shared with anyone who asks. Analysts rated this settlement as a "slap on the wrist" and noted it was set to expire in 2007 anyway. One interesting argument in Microsoft's appeal was that breaking the company up would make it weaker in the international market. This is another noteworthy detail, as it shows these companies know they're being used as global economic weapons. EU wasn't so considerate of Microsoft's overseas exploits, though.

In 2004, EU and its numerous commissions agreed that Microsoft was indeed stifling competition and slapped it with a $794mm fine, which it begrudgingly paid. In 2006, there was another $448mm fine and in 2008 a whopping $1.4bn fine, all of which Microsoft appealed until all avenues were exhausted. In the end, EU representatives admitted fines might not be enough, as it seemed nothing force Microsoft to change its ways. Instead, the preferred way would be settlements wherein the company would agree to actually do what it was asked to. The pragmatic answer would be that Microsoft still makes so much money through its monopoly that fines are worth it.

In 2009, Microsoft was compelled by EU to provide a choice immediately upon Windows installation of one or more of 12 internet browsers available at that time, including Firefox, Chrome and Opera. The problem was that Internet Explorer was baked into Windows, causing problems to developers of other browsers who had to accommodate Windows rather than their users. Microsoft eventually stopped offering this choice and EU issued a $731mm fine, give or take, in 2013 for that,. Windows Media Player was another point of contention for EU regulators, so Microsoft was eventually tasked with producing a version of Windows without it; nobody really cared about buying that version.

Consider the egos of such men that a $1.4bn fine does nothing to change their minds. Everything you see in Silicon Valley about CEOs of tech giants is the total and absolute truth – these people see themselves as Messiahs chosen to lead the entire world to enlightenment through semi-useful products and services. Bill Gates even announced a lofty goal of eradicating diseases[17], such as smallpox, from the face of the Earth through aggressive vaccination and better sanitation.

In both instances, with Internet Explorer and Windows Media Player, EU legislators came to a correct conclusion—these two programs were offered as defaults and blocked development of alternatives. But, what's the big deal with default programs? They exploit human inertia in a very insidious way, lulling us into complacency. One of the first tasks you'll have as an ethical hacker is to choose your own suite of tools that work just the way you like them and for a good reason—defaults suck. Most people suffer in silence and just use whatever's provided as default, but you'll have to do better than that.

GPL or GNU Public License[18] is what's known as copyleft, an alternative to copyright that's being used by established tech companies to deny upstart competition access to the market. Under GPL, you get truly free software in the sense that it costs nothing, and you have the freedom to do with it as you please. Ideally, we'd all slowly transition to open source software, such as by using Libre Office instead of Microsoft Office and Linux instead of Windows.

Chapter 6: The Danger Of Defaults

Understanding default settings and the danger they present means understanding how the human brain works and why it works that way. The human brain is a living optimization machine, constantly trying to find shortcuts to save time and energy. If there's a beaten path in front of us, why wouldn't we use it? The problem is that we might become so efficient at following the same path even when it becomes dangerous to do so, that we might ignore the hidden gems under our feet or details that warn of danger or indicate better paths.

The same efficiency with which the brain eases itself into a routine ultimately makes us complacent and inert. Without a corrective influence or an alternative, the groove becomes self-defeating. At that point, we start opting for comfort and known solutions, even when they no longer work as they should or fail to work at all. This can be marked by the frustration and anguish that commonly arises when using dysfunctional and obsolete tools and methods or by a sense of euphoria and invincibility that's common before we smash face first into reality.

Animals use the same brain pathways when they get domesticated, which is why a stray cat you've fed a couple times will start rubbing against your shins. It's easier to cajole this big sucker into dishing out a can of tuna than rummage through trash for a snack. We all relish finding a good shortcut, and there's nothing wrong with doing things faster as long as there's nobody trying to abuse our sense of efficiency. Digital brains that power tech giants' algorithms are designed to work like living brains, meaning they exhibit all the same optimization-seeking behavior and weaknesses we do, but with one exception – we can adapt beyond initial conditions through curiosity that motivates us to learn from others in a way no machine can.

A grave weakness of humans is that we assign our qualities to things, thinking they must be human if they show a shred of intelligence. In this case, this means seeing Google's digital brain at work and thinking "there must be a human on the other side".

It's not that there's a wizard behind the curtain but that there's nobody back there. It's all just a ruse, a machine made to mimic human behavior and left in charge of things to cut down on hiring, training and paying actual people. Once you realize that code is law and that the machine is extremely optimized to do a limited number of things, it becomes quite a simple adversary.

Slick design is another cozy trap for the brain—you're using smartphones because the corners are rounded and are simply begging to be touched, the icons are so cool, and animations look great. These features are mesmerizing on purpose and are meant to ease you into repeating the same meaningless action over and over again. In short, you fall in love with the task but not the result it's meant to produce. If you can ignore the appearance of products while focusing on their essence, you'll start thinking like an engineer.

The amazing part is just how adaptable our brain is. Throughout life, we'll be constantly learning new things and skills as long as we're willing to try them. Having that potential wasted on a hamster wheel product is the real tragedy of the modern age--the fact we're not being challenged to become better. It seems like there's nothing that can replicate the same thrill of discovering a new thing as a kid, but hacking begs to differ. Life is nothing more than a game, only as an adult, the toys you're playing with cost a bit more.

Tech giants still wield massive influence, namely through media and legal departments that can wreck your reputation and livelihood, but as long as you remain on the ethical side of hacking and don't expose yourself too much, you'll do just fine. When you start playing with tech giants' toys in a way they never intended, you'll discover just how slipshod they are, and at that point, you may do as you please. Your greatest asset is your curiosity, which you can even monetize by offering courses, lessons and e-books based on what you've learned about ethical hacking. But as mentioned before, make sure you stay within what's legal to avoid trouble.

Chapter 7: John Deere

John Deere, the tractor manufacturer, also started incorporating EULAs into the sale of their heavy machinery[19]. Their justification was that their tractor contains a computer with proprietary code that might be accessed, tweaked, improved by customers or used to pirate music, so a EULA was needed to give legal leverage to the company. If a tractor model involves IoT capability, John Deere might also have access to telemetry and be able to shut down the machine remotely for no reason. So, farmers became ethical hackers.

By buying a dummy tractor part for $25, a Motherboard reporter got access to a private Ukrainian forum where cracks and tools for John Deere tractors are sold[20]. In essence, this software allows a John Deere tractor-user to become their own authorized repair shop, fixing things on the spot with second-hand parts. Cracks can also overclock the engine, remove restrictions and make the tractor work as expected. What's interesting is that in 2015, Librarian of Congress made an exemption to copyright that applies to land vehicles, such as tractors, exactly because of the right to repair. John Deere then required tractor buyers to sign EULAs that can be used to invoke breach of contract rather than breach of copyright.

Farmers then started the Right to Repair movement, with legislation welling up in states across the US and farmer organizations endorsing the movement and bills. The logical notion of actually owning and repairing the device you bought didn't sit well with John Deere, who claimed[21] it created less-than-optimal customer experience and the free market was good enough to address the issues. Especially ironic is the plea to keep the government out of private contracts formed between John Deere and their customers. Apple also interfered with second-hand repairs of iDevices, claiming that people repairing their own smartphone would lead to broken glass which can cut fingers. As we said previously, companies consider their users idiots and intentionally make things that drive smart ones away.

Why doesn't the free market just offer an alternative? It's obvious that John Deere customers would gladly take a vehicle that lets them upgrade and repair it themselves; how come that hasn't happened yet? Thanks to the outdated notion of patents, established companies can leverage their existing businesses into a technical monopoly that doesn't attract the attention of the government or break any laws.

It doesn't matter if the competitor actually infringed on any John Deere patents so as long as there's a shred of possibility the giant company will use legal processes to keep the competition out of the market. An upcoming competitor can be sued into bankruptcy because John Deere is already selling products and has the revenue stream to support legal action, which the competitor can't afford to fight against.

So, where does that leave EULAs? They're flimsy but companies don't need much to act on. EULAs merely provide the perfect catch-all excuse, at which point it comes down to a war of attrition and finding the court to agree with you. Companies will always have deeper coffers than individuals, so unless you can find someone to fund your cause the same way Hulk Hogan got Peter Thiel's backing to bring down Gawker[22], your chances of winning are slim to none. You shouldn't bother reading EULAs because they can change without prior notice and the clauses in them apply anyway, unless they are utterly deranged. The only real solution is to try and use open source software, free alternatives to these EULA-laden travesties sold on the market.

Chapter 8: Youtube Content Id

On YouTube, Google uses a patented content analysis system called ContentID[25], which compares any video and audio uploaded against a database of copyrighted works. Which authors get the privilege of being protected by ContentID? Only those that "own exclusive rights to a substantial body of original material that is frequently uploaded by the YouTube user community," meaning Disney, Nintendo, Sony and other big companies. Meanwhile, the little guy uploading videos of his cat to YouTube has to tremble in his boots expecting someone will issue a copyright claim on the basis of something in the title or background audio.

In 2009, Scott Smitelli got an automated email from YouTube informing him that his sleepover video where three of his friends dance to "I Know What Boys Like" by The Waitresses was removed. It's when his car commercial parody video was also removed due to audio copyright issues that he got intrigued. Neither video had nearly enough views to attract attention, and the content was fairly obscure even in its heyday. It wasn't likely a human filed those copyright claims, so who did it? Or better yet, what did it?

Scott decided to run a series of tests[26] to discover how ContentID works using that same 1982 song, which he happened to have in an uncompressed format, by playing it in reverse, altering pitch, manipulating time compression, resampling, adding white noise, recording the song with a camcorder in a different room, changing volume, splitting the song to pieces and messing with stereo channels and vocals. After uploading a total of 72 videos, Scott was impressed with how robust the ContentID system is.

ContentID can detect copyrighted audio even when the volume is lowered down to a barely audible whisper and can also hear through up to 50% white noise which would make the audio useless to humans anyway. Fooling ContentID meant altering the audio so much that humans couldn't recognize or enjoy it anyway. But there was one trick that worked—cutting out the first 30 seconds of the song consistently defeated

ContentID, presumably because that's the only part of the audio it checked. Scott did his tests in 2009, and it's likely that ContentID was improved since then but the scale of Google's automation makes the system fairly defeatable if you're not afraid to experiment. For example, copyrighted video can slip by if you change the aspect ratio or rotate the output.

Copyright claims are an issue to which nobody has a solution and which Google has simply offloaded onto users. To be fair, YouTube may allow filing a counterclaim, but that has to be done manually by the affected channel's owner. In cases where the YouTube video in question is monetized and makes money showing ads, whoever files the copyright claim can be awarded ad revenue the video generates until the counterclaim is filed and resolved[27]. Enter YouTube copyright trolls.

When the owner of YouTube channel, "Dope or Nope", realized the problem with copyrighted works, he decided to create a separate series called "Lyrics in Real Life", in which he would take parts of lyrics from popular songs and write a dialog around them. Almost instantly all those videos were claimed[28] by random third parties, regardless of whether they qualified for the fair use exemption. Another YouTube creator, Gus Johnson, also had his videos claimed by a song owner because he was playing the song using a paper towel dispenser[29].

Anything and everything on YouTube can be claimed and building a channel without using copyrighted work is impossible. "Dope or Nope" owner even retells how two of his videos got claimed by a music studio because they featured a couple artists employed by the studio. That's it, without using any copyrighted audio, lyrics or titles, the two videos apparently infringed on the studio's copyright. But there's no way to tell how because whoever files the copyright claim isn't expected to explain any of it when filing. Even if you use a fraction of copyrighted work, the copyright holder can lay claim to your entire work and make money off of it by showing ads.

Filing a YouTube copyright claim[30] is as easy as visiting the help page and clicking through a couple buttons to detail what the issue is about. The options provided include "someone is abusing me," "I see nudity or violence," and "someone is using my image," which tells you all you need to know about how DMCA Title II is merely being used as a catch-all tool for censorship and harassment. Google does threaten to punish those who abuse this process, mainly by shutting down their account. Oh dear, what will they do then? Google also mentions the possibility of legally pursuing the same, however to date, there hasn't been a single instance of this occurring.

Seeing how Google has no support staff at all servicing regular users, your only recourse is to plead on unofficial forums[31] where Google sycophants will readily answer all your questions with boilerplate answers. Good luck trying to get an official reply. Those are reserved for enterprises that actually give money to Google, not just some ordinary plebeian who can't get his videos monetized for over a year.

Worse still, Google is trying to turn YouTube into a regular cable TV experience, despite the "you" in the name. So far, all attempts to show regular programming on YouTube have failed. YouTube Red is the prime example, but this shows how regular users are intentionally being pushed off the platform. Ironically, the idea of making the world a better place by offering a worldwide streaming platform is hampered by copyright laws and authors who would have to agree to less-than-ideal conditions to have their work shown in, say, South Africa.

Even when you respond to a YouTube copyright claim, there's no person at Google that reads it, and your response goes straight to the person who claimed it. Anecdotal evidence shows a YouTube user playing classical music on a piano has a roughly 50-50 chance of getting his videos claimed by a major music publisher as copyrighted[32] and a wedding video with music playing in the background can get claimed just as easily[33]. As time goes on, the number and array of copyright claims increases for no apparent reason. Even if you decide to go out into the wild to reconnect with nature, absorb the good vibes and get original content for YouTube, your videos can get claimed—a video containing no other sounds than 50 hours of rain falling got claimed by 5 different entities[34].

What happens with the claimed video is at the discretion of claimant. ContentID might result in a video showing ads, being blocked in certain countries, having its audio muted or being taken down completely. One method of dealing with this is to lay a copyright claim on your video before anyone else, but Jim Sterling recommends using his "copyright deadlock"[35] method to make sure nobody gets money off of your work.

After having one of his videos claimed by three different video game publishers, Jim realized that one of them wanted the video monetized while the other two didn't. This resulted in a conflict that Google's algorithm didn't know how to solve and therefore just left the video unmonetized. So, instead of scrubbing all traces of copyrighted work from your YouTube videos, you should intentionally aim to infringe on it to just the right degree.

Because big companies, and Google is one of these, are too busy to hassle with individual claims, they'll leave ContentID running and just check up on it once in a while, perhaps tightening the screws if they figure out they'll get away with it. By learning how it works and how to exploit its shortcomings, people can often do as they please on YouTube without any fear of repercussion. The same applies to all platforms, though the fact YouTube deals with thousands of hours of new video each second does make people's ethical hacking easier.

This poses an interesting question regarding a future in which we're surrounded with algorithms such as ContentID working in favor of big companies. Should we even rebel when we can find a loophole so easily? Ethical hackers don't even register on the radar, and they'll always be an insignificant minority of users because these products are meant to make us complacent. By realizing how the companies think, ethical hackers can make them work in their favor, and nobody will be the wiser as long as they're ethical about it. An interesting part is that companies couldn't change their ways even if they wanted to. Shareholders demand maximum efficiency, which means ignoring what seems like a statistical error-- a small contingent of users that somehow defies all rules and restrictions.

Chapter 9: Tracking Users

But what is the motivation for Google to keep offering YouTube and other services to regular people? As it turns out, the real motive is profit. Google can afford to give out free space and bandwidth because by requiring users to make an account, they get access to private information, such as phone numbers, that can be used in data mining. Yes, your phone number is that valuable. There's also a chance that users will start relying on Google for business needs, which makes them locked into the Google ecosystem of products and services. But the real source of revenue for Google and other tech companies is advertisements.

In short, all tech companies race to create the perfect product that will serve as a funnel for leeching data from users and in turn, display appropriate ads. Instagram, Viber, Facebook, Snapchat, YouTube and even Windows have become spy tools that do provide some utility but also spy on the user and display ads. And it's all done legally because the user consents. Just take a look at the list of permissions any social media app requests before you can install it on your smartphone: camera, microphone, storage, texts, calls, address book, location, Wi-Fi network and so on.

By combining those tidbits of data with what the user shares of his own volition through messages, status updates and uploaded content, the company running a social media app can create a comprehensive profile of each user and make money selling it off in some capacity. Even if you decide you won't be using any of those apps or own a smartphone, your presence is tracked and analyzed in some way, such as through people in your life who do use smartphones and social media apps. Offline activity data is gathered as well and has been gathered for decades from sources such as store loyalty cards.

Profile creation is done by using smart machines that have been kept in the dark and allowed to compete against one another, resulting in evolved software that's highly specialized in a certain area, such as recognizing objects in an image. So, this kind of software can scan your selfies to see who's in the picture, where it was taken, the time of day, location, objects etc. in an instant. In this way, companies can triple dip: sell user profiles, serve ads, and perfect the smart machines, all the while honestly claiming that privacy isn't being jeopardized. You see, it's the machines that do the spying. There's artificial intelligence all around us, but the companies can't allow us to interact with them freely or even know about their existence because—we'd mess with them.

By uploading fake information, taking fake selfies or just constantly lying about everything online, you're messing with the artificial intelligence and corrupting the data the company is gathering on you and the demographic you belong to. Because even the smartest machine is kept in the dark and fed user data, it can be made to spit out wrong information, tanking the created profile value. After that, you can find ways to block ads whenever you're surfing and strip intrusive apps from your smartphone, if you happen to be using one, thus denying any company all streams of income from your online and offline activity. Then you can start doing some hacktivism, political hacking.

Mainstream media would want us to believe that hacktivism has to do with intrusive, obnoxious and disruptive hacking, such as breaching a company's servers to uncover corruption. Hacktivism is nothing of the sort but rather a low-key, low-risk hacking of the political and psychological process to create a change in the general mindset of people around you. Thanks to the butterfly effect, we know that small actions done today have massive consequences years or decades in the future. So, rather than becoming a highly visible politician that attracts unwanted attention, you can effect change by simply mentioning how there's an alternative to Google and that online ads can be blocked. If enough people were to do this kind of hacktivism, the tech giants would experience a phantom decrease in users and revenue that couldn't be traced back to you or any other single source.

Social media networks are a special kind of evil in that they feed on the most powerful drive humans have—social proof. For the same reason encountering a group of people that look up will give you an irresistible urge to also look up, social media acts on your instinct to stay in touch with reality by observing what people around you are doing. Studies have shown that social media networks tend to cause actual depression to frequent users[36], primarily by making them feel like their lives are miserable in comparison to others. If you still can't stop using social media, start lying about what you're doing. This will give you an enormous sense of freedom in that you'll no longer feel obligated to keep up with the Joneses. Again, if enough users were to lie on their social media profiles, those companies would all of a sudden find their value tanking for no discernible reason.

An engineer would only care for your metadata, meaning how many times you've started an app, when it crashed, why and so on, but businessmen and marketers would like to know everything and really have no idea what to do with that data except sell it to other companies. When these streams of private and metadata keep pouring in, the company in question usually starts skimping on storage and security, meaning hackers have an easy way of getting into these servers and scooping up whatever they want. It's not a matter of "when", but of "how often" and "how long".

Facebook earned over $13bn in just three months in 2018, most of it coming from ads. With over 2 billion users, Facebook is a juicy target for all sorts of scammers and hackers, who have no trouble attacking company servers due to poor staff training and a large attack surface. Mark Zuckerberg has already appeared plenty of times in front of Congress related to privacy and hacking scandals, explaining away both as an unavoidable consequence of servicing the entire world. Now he wants to engage in politics, somehow hoping he'll make the world a better place.

The real solution is to surgically remove Google and its tentacles from your life, meaning to degoogle[37]. By finding alternatives to Google products, even if it just means using something other than the Google Calendar, you're making yourself almost like a ghost online: unseen, untraceable and unfathomable. Remember not to reuse passwords and ideally have a separate email address for each new service or product, lessening the chain-reaction impact of any hack. If you're offered the option to use security questions, make up false but memorable answers. In this way, even those that know you intimately won't be able to gain access to your accounts. Whenever possible, diversify your online existence and fragment it as much as possible, keeping truth only for yourself and those you trust, rather than anyone with internet access.

Cookies are small text files set on your device whenever you visit a website. Their legitimate use is that they help the website recognize a returning visitor and, for example, log you in automatically. The problem is that websites are no longer isolated entities with content you see originating from one source, so cookies can be set from a completely unexpected source without your approval or knowledge. These are known as "third-party cookies" and if you just briefly go through any privacy policy of any website, you'll quickly notice they're constantly being justified as necessary. They might be necessary to the website and the data gathering entity that funds the website, but there's nothing in it for the user.

Think about what kind of a rotten deal this is--you get tracked and have to bear ads in exchange for content that might as well be free. So each user creates a unique stockpile of third-party cookies that identifies where he's been throughout months and years of using the internet. It's all a chaotic mixture of dozens or hundreds of websites that share code, content and information amongst themselves, creating a web of surveillance. It's hard to say whether any of it was intended but here we are, and countries are starting to take notice.

Cookies became such a big issue that EU actually created a massive legislative barrier called GDPR and enacted it in May 2018. It was meant to tilt the scales in favor of EU-based companies. Well, it did nothing. Websites know barely anyone reads privacy policies or even knows what cookies are meant to do, so nothing of substance has changed. Again, you can see legislators trying to cover up for dumb customers, but nothing helps except becoming adept at ethical hacking, meaning adapting to the situation by tweaking hardware and software to suit your needs.

Turning off cookies might mean websites won't work as they should, and, in some cases, they might even deny you access with the warning message saying as much. Simply put, there's no expectation of privacy online, and those who want some anonymity are cut off at the pass. Yet, you can achieve digital privacy to an extent by blocking third-party cookies, the real method of tracking. You can also use private browsing modes, which all major browsers support, and after which, all cookies and browsing history are cleared. In private browsing mode, a website sees you as a blip on the radar and then you're gone for good. If cookies are allowed to pile up, it's like you're wearing a GPS bracelet that lets someone track your movements to figure your behavior.

Mozilla Firefox started a trend of letting users opt out of being served third-party cookies through a feature known as Do Not Track. The problem is that websites would have to cooperate and voluntarily abide to not serving tracking cookies, which impacts their bottom line. As expected, barely any websites agreed to not tracking users. So, in August 2018, Mozilla Firefox started aggressively blocking third-party cookies and made Do Not Track an ultimatum rather than a polite request[38]. It's a noble attempt but no company can cover your tracks for you.

When browsing the internet, use an adblocker to stop the leakage of private information and squelch intrusive ads that distract and clog up your device. All major browsers should have an adblock extension available at the extension store, and there are even browsers on mobile devices that have an adblock built in. Do keep in mind that companies serving ads consider them an essential part of the browsing experience, so your adblock is essentially hacking the service. While we're at it, why not change the entire page to suit your liking?

Right click on any content inside your desktop browser and use the option "Inspect element" to tinker with the source code of the website. This will usually open a tool known as Developer Console where you can see the code tick in front of your eyes as you use the website. Click through different parts of the code, copy it to Notepad, tinker with it, paste it back and see how the website changes. You are the owner of the code displayed in the Developer Console and may do with it as you please. Do note that reloading the page will reset all your changes.

Experiment with different options and get acquainted with what classes and elements do, because an adblock typically allows blocking individual elements on a page if you know its source code. See a picture, toolbar or button that's annoying you? There is a way to block it for good, though it will take some practice. All adblocks typically draw inspiration from the granddaddy Adblock Plus, which has the most extensive help, in particular on how to block specific elements[39]. Experiencing this kind of raw power, the kind that lets you customize your browsing, is truly liberating and is perhaps the noblest kind of hacking, as it teaches you about the software you're using in a constructive and practical manner.

Websites have tried introducing anti-adblock scripts that detect adblock users and prevent them from viewing the content, but that can be bypassed too, either through dedicated anti-anti-adblock scripts or by simply disabling Javascript, dynamic programming language. Developer Console might have an option buried inside its settings that allows for disabling Javascript, alongside a few other surprises. Even disabling Javascript has now become a problem, since in November 2018, Google started requiring that you have Javascript turned on[40] or you won't be able to log into your Google account.

Some hackers have created a magnificent Firefox addon called AdNauseam[41]. Translated from Latin, the name reads "until one is sick of it" and involves blocking ads and clicking on them for you. What the advertisers are actually after is creating the perfect ad that is so in tune with who you are and is presented at just the right moment, so you have to click on it and buy the product. AdNauseam corrupts whatever behavior profile the ad agencies have created by clicking all ads without requiring any input or resources from you, which also ruins the rate at which the website gains money from ads.

Ideally, you would have several browsers on your machine and use one for services that require a login, another for casual browsing and third yet for watching content you'd rather others don't know about. By segmenting your online personality between several browsers, some of which can have cookies, Javascript or ads blocked, you reduce your attack surface and make it difficult to track you across the web. There is still the matter of concealing your IP address, a virtual home address that can be used to trace you back to your country, ISP, city and street address.

By using VPN, virtual private network, you can go through a device that's physically located in another country or perhaps on another continent, gaining its IP address. There are hundreds of VPN services[42], which is great because you can find a small VPN company that will help you hide your tracks without adding your data to the tech giant portfolio. Of course, law enforcement can subpoena VPN companies to get access to user data but then you'll be having much bigger problems than worrying about IP addresses and cookies. If you don't want to shell out cash for a VPN, you can try using Tor, a free VPN.

Tor is an open source software VPN and, like all open source solutions, all but requires that you understand and love hacking. By bouncing internet traffic between all Tor users, this VPN can mask the origin of any user and let them do things such as whistleblowing or reporting from dangerous places. The name refers to the idea that trying to look for any given user in a Tor network is like trying to find the center of an onion by peeling it – it's layers all the way down. Do keep in mind that, while the traffic is encrypted and tumbled inside the Tor network, the exit nodes are likely surveilled by CIA, NSA and other such agencies looking for users who changed any of the default Tor settings to track them by. You're playing with the big boys now, so stay safe.

Websites can also collude to use what are known as "tracking beacons", which are pixel-sized images that are there simply to track a user across websites. By visiting a number of pages loaded with tracking beacons, all websites involved get a comprehensive behavior profile of the user--which pages were visited, what was clicked, how long the stay was and so on. All major browsers should have an extension that helps with blocking these tracking beacons, but the best solution is to identify these sites and not visit them at all. If you can't find one, you might want to learn how to make one yourself.

Chapter 10: Drm

DRM, or digital rights management, is a set of tools used to enforce any given EULA and can involve hardware and software checks to make sure you're not coloring outside the given lines. Entering serial numbers, calling tech support to have your copy activated or simply having to be connected to the internet at all times to merely use the product are all DRM. Tech companies love DRM because it gives them leverage over clients' businesses that can be used to suck their wallets dry. DRM also provides an insight into customers' behavior through data mining, surreptitious gathering of personal data and metadata, usage statistics.

Cloud refers to remote storage and is DRM as well, arguably the only DRM done right as there's actually some utility to the customer. By having devices that store the majority of content to company's servers, users don't have to worry about buying extra storage with the device or updating programs. The company also gets an excuse for device requiring an internet connection while gathering metadata.

DRM is particularly brutal with video games, hurting and annoying legitimate users. The idea for DRM in video games is simply to block reselling, sharing or copying games but some methods are so Draconian that they can actually ruin a machine they get their hooks in. StarForce is a notorious example[43], installing shoddily-made drivers that can prevent Windows 7 and 10 machines from even starting up, and nothing short of operating system reinstall helps. Keep in mind, this is software made by international companies worth billions and legitimately bought by dumb, honest customers. Ubisoft was the most prolific user of StarForce until they got sued for it[44].

Those who take a course in ethical hacking can engage in cracking, which is about removing DRM and simply making software freely usable without any restrictions. Crackers typically do it for bragging rights, with some video game developers being so lax about DRM that they release preorder copies with it, allowing crackers to defeat DRM before the game's actual release[45]. It's pretty tragic that users who program for free can easily defeat corporations that are supposed to create a workable product that will go on the market. Cracks are used by simply replacing the DRM-laden executable and other files with cracked ones that don't have it. So, why don't companies make their software uncrackable? You already know the answer—it would cost so much money, and it would still be defeated in the end that it's not worth it.

Ubisoft was the company that released Rainbow Six Vegas 2, which had a DRM so buggy that it checked for the presence of a non-existent CD in the drive after the game was patched. So, legitimate buyers couldn't even play the latest update, but what about pirates, those who ignore copyright and freely share software? They got rid of DRM and essentially fixed the bug, releasing a more functional copy of the game to their fellow pirates. In the end, Ubisoft borrowed the crack made by pirates and released it officially[46] when enough customers started complaining. In a twist of irony, the multibillion-dollar corporation broke copyright by taking someone else's code and profiting off of it.

In the past, DRM was at least trying to be unique[47]. Feelies were physical objects that were meant to prevent the copying of access codes and serial keys that would normally come with a video game. Today, you'll get statues, dice, playing cards and whatnot for buying a deluxe boxed copy of the game but in the early days of video game production, those were the DRM. The idea was that security keys printed on a piece of paper could be copied and quickly shared with friends alongside the copy of the game, but if a statue, map or plush doll was used for DRM, then it would supposedly be so difficult that it would discourage potential sharers.

Lenslok was a foldout series of prisms that had to be put against the monitor while the game was running. The prism would unscramble gibberish text on the screen to provide access codes to advance the game. Of course, if you had an unusual aspect ratio, quirky monitor resolution or just a big TV, the prism wouldn't work, and you'd be stuck with a botched game. Leisure Suit Larry was a raunchy video game series that used a DRM with pictures of women and their phone numbers (we don't dare guess how many work hours were wasted trying to guess the numbers).

Other video games came with manuals that were chock full of pictures, maps, diagrams and riddles. At any time, the game could pop up a prompt and check if your were the legitimate owner. You can't type in the 6th word in the 2nd paragraph on page 25? Tough luck--the game locks up and won't let you move past the check. Console cartridges were a DRM in their own right, making it difficult to copy games. For Super Nintendo's Earthbound, trying to play the game on anything other than the original hardware would let the player advance all the way to the end before freezing and deleting all saved games. Arma 2 had what's known as FADE copyright protection, which would slowly degrade game performance on an unauthorized copy[48].

SecuROM was another botched DRM scheme in the vein of StarForce. Developed by Sony and used by companies such as EA (Electronic Arts), SecuROM would install itself and contact company servers over the internet without giving off any signs of presence until the same game key was used three times. At that point, SecuROM would kick in and block reinstallation, remaining on the machine even after the game was uninstalled for good. Users could phone tech support to have their activation limit lifted, except that in the case of BioShock, the phone number printed on the manual was wrong. After EA got hit with a lawsuit[49] over SecuROM, this nasty malware-like DRM scheme was retooled to allow reclaiming installations and included a removal tool. The only way to remove it prior to that was reinstalling the operating system.

GOG.com is a big middle finger to DRM and all companies using it. Being a licensed distributor of video games, mainly abandoned and obscure ones, GOG (Good Old Games) bought off rights for video game classics, ripped the DRM out and started selling them. That business model of simply putting out a video game that doesn't ruin your machine was received favorably by players and GOG's portfolio increases on a daily basis.

Xbox 360 was gutted by the use of DRM that required licensing games and having an internet connection to play them. So, you'd license a game and not even own it, but you could at least play it, right? Thanks to Xbox 360's infamous Red Ring of Death, complete hardware failure caused by overheating that weakened newly introduced lead-free solder meant to comply with EU eco-friendly standards, consoles were constantly dying on users. All games that were first downloaded on a console that died were only playable as demo versions. A Microsoft insider revealed about 30% of all Xbox 360 units experienced RRoD[50] and an interview with Peter Moore, then-executive at Microsoft, revealed the recall cost the company $1.15bn[51].

Steam is an online marketplace for video games that's also used as DRM, forcing you to stick with an account that can be shut down at any moment with all your investment gone in an instant. The difference between Steam and SecuROM is that Steam at least works as advertised. Steam also has additional benefits, such as sales and gifting games to others. If there

must be DRM, then that's the kind of
functional DRM we should be getting.

Chapter 11: Torrents

Torrent is a distributed, peer-to-peer sharing network that bypasses search engines and tech corporations to create efficient ways of downloading data. For example, BitTorrent is one torrent client, with its protocol specifications page[57] going into greater detail on how it works. In a torrent network, "seeder" is the person that initially shares the file and "leechers" or "peers" are the ones downloading. When peers get the whole file, they become seeders too, speeding up the network. A massive advantage of torrent clients is the ability to cap download and upload speeds, so your internet connection doesn't choke but they also allow you to continue downloading whenever you quit the client.

Torrents are legally dubious and although you can upload your own files to share, an overwhelming majority of people uses torrenting to share copyrighted material, such as movies, video games, comic books, magazines etc. The trick is that downloading a torrent can't begin unless the owner shares the hash, a mathematical summary of his file, either by sharing the torrent file or what's known as magnet link. Don't worry, it's all pretty simple once you download your first file through a torrent client. So, the main hurdle is finding a reliable way to get access to as many torrent file hashes and magnet links as possible. Enter Pirate Bay.

Also known as "the galaxy's most resilient BitTorrent site," Pirate Bay has been championing the right to freely use copyrighted works and getting its share of flak from governments and private copyright enforcement bodies. Be warned; this is what happens when you go into the spotlight challenging copyright–you'll be attacked from all sides so your best bet is to just stay low.

Started in 2003 by Swedish opponents of oppressive copyright lobbies, Pirate Bay serves as an index for torrent files, freely accessible and searchable by anyone. So, how come it hasn't been shut down already? The thing is, Pirate Bay has been shut down hundreds of times. After each time, it comes back stronger and even more defiant, as if to spite the copyright enforcement authorities. It's pretty interesting to watch the mightiest copyright agencies and lobbying bodies playing a game of whack-a-mole with a simple website that's basically just a database of hashes.

The three founders of Pirate Bay will later be arrested and sentenced to a year in prison plus a fine. In 2006, inspired by Pirate Bay, Swedish Pirate Party will appear led by Rick Falkvinge, with the idea of reforming copyright laws and helping usher copyright into the digital age. It will become the third largest party in Sweden, grabbing the attention of youth but otherwise remaining neutral on all political issues. Pirate Party and its swelling numbers will persuade other parties to soften their stances on copyright laws, though it's curious that only pirates question the idea of people who simply share files being jailed for it.

QBittorrent is a great torrent client with a side magic of Python. Rather than risking exposure by visiting Pirate Bay or any other website hosting torrent files and magnet links, you can install qBittorrent5 and

Python to enable a powerful search engine that can use plugins downloaded from GitHub and other places to directly search for whatever torrents you want. Don't fret, it's really easy and opens up a whole new world of content that's undetectable from the outside. If you get stuck, just ask qBittorrent community online and you're sure to get plenty of pointers.

So, why don't ISPs block torrent traffic? It turns out they have more than enough on their plates, having to deal with hundreds of gigabits of data per second. There's no possible way to scan even a fraction of it in real time, let alone when it comes to encrypted traffic. What ISPs do is take up individual requests from users, put them up to other ISPs globally and let everyone pick whatever traffic they think they can serve. This is called Border Gateway Protocol and essentially represents a web of trust, with numerous organizations sifting through the traffic data afterwards to find out who acted as a good guy and who acted as a villain.

Some lawmakers actually tried to put the burden on ISPs to scan their traffic for copyright infringement, but all those proposals were quickly shot down as unfeasible. An ISP would have to shut down its entire infrastructure for a month or so to implement such fundamental changes, at which point other ISPs who don't want to monitor their users would simply stay up, hoover all the available users and move on. Internet is a rare case of the free market protecting fundamental human rights and working to foster freedom of speech, interaction and transfer of goods and services. We'll likely never again have this freedom of speech and sharing, so use it right and don't let yourself spend your life glued to a social network.

In this tornado of data, your digital requests will be just a whisper in the wind among millions of other whispers, insignificant but ultimately detectable and traceable. As long as you keep your digital identity as fragmented and as detached as possible from your real identity, you'll be like a ghost moving through the ether, safe from grubby hands of braindead legislators. This kind of genuine anonymity is intoxicating and can help you understand why people become black hats in the first place.

You can abuse freedoms you now realize you have but law enforcement will eventually catch up to you, whether it's 5, 10 or 15 years down the line. However, you can also use these freedoms sparingly and for good causes, becoming an ethical hacker, one who understands how the levers of technological power can get stuck and need just a tiny push to get them moving again. What you decide is ultimately up to you but there are plenty of common people being terrorized by government-endorsed villains who could use your help.

One easy way GEMA detects pirates is by setting up a honeypot, an area of intense scrutiny, at a website sharing movies, videos, music or torrent files. By simply connecting to that website from a German IP, you've left a definite trace of your intentions, and that's all GEMA needs for a court order to barge into your home and harass you. However, that same movie, video or music shared via a torrent file or a magnet link lowers the chance of being detected, and simple, inconspicuous encryption lowers the chance even more. A magnet link is just a short snippet of gibberish text that can be hidden inside a text document, lowering the chance of being discovered even further.

There are no certainties when it comes to copyright and torrenting, only layers of protection. It's tragic that it's come to this, but artists in Germany and those wishing to share their works with Germans should engage in torrenting, a decentralized and secretive form of sharing that causes more headaches for GEMA than any other form of public performance liable to fees and penalties.

Is there anything stopping users from sharing malware disguised as movies or cracks in torrents? Nope, but you can keep an eye on strange signs, such as a low number of seeders, which would mean people have figured out it's not the real file. In short, torrenting is low-risk in one way and risky in another, but it's all down to your experience on where to find the best torrents and how to spot fake or dead ones.

The best way to use torrent clients is on a separate machine with an antivirus and a firewall for just-in-case. If some malware does get through, it's better not to have anything important on that machine.

Chapter 12: Sports Channels

How about watching sports? It seems no cable TV package has the option to watch everything, especially if you savor some obscure sports, such as Polish women's handball. Luckily, your friendly pirates have your back, providing you with more sports channels than you can throw a computer at. Cable companies would like to have you believe they are the ultimate arbiters of who can watch what and where, charging you an arm and a leg for the privilege. It would be fair if the corporations involved with streaming rights could just be honest, but they can't resist trying to finagle their dumb customers out of an extra penny.

US sports in particular are subject to some idiotic restrictions that make it difficult to find out how many matches you're entitled to see for the price, with NFL deservedly getting the most blame. NFL is the name of the organization that owns the US brand of football, that being hand-egg or gridiron football. Yes, they own all copyrights on team names, emblems, colors, player names and so on, with the end result much worse than you could ever imagine. All NFL matches consist of live streams, in-game replays and post-match analyses, which are distributed to select companies.

Soon after its creation, the NFL came up with a strategy of regional blackouts, which means that a match can't be shown live on television stations within a 75-mile radius of its stadium in order to promote ticket sales. Prior to 1973, the blackout rule covered all home matches under all circumstances. Bedridden and want to watch your favorite NFL team? The NFL doesn't care; get up and drive 75 miles to the stadium in person. It took presidential intervention and an explicit threat of anti-trust legislation specifically aimed at the NFL in front of Congress before the blackout rule was loosened. If at least 85% of all stadium tickets have been sold 3 days prior to the match, the match can be shown in the region.

Try to wrap your head around this reasoning—if the tickets for a match are sold out, that means there's interest in the match, so local TV stations can air the match. If tickets weren't sold out, there's no interest in the match so the rights to air the match can be sold to someone else. But, that's not how sports work. The blackout rule completely ignores human psychology-- how people watch and enjoy sports and, in particular, how people become sports fans and what effect they have on the regional market.

A sports fan would want to support the team when they're going through a tough period and will read about them in the newspapers, buy their kitschy merchandise, watch the TV broadcast or buy the ticket to shout and flail from the bleachers in inclement weather. So, a fan of a team isn't affected by the blackout rule but those who are mildly curious are effectively stopped from even finding out what the sport is about. With the blackout rule, the NFL literally censors the match out of the public consciousness and doesn't let the public at large, especially children, fall in love with the excitement of gridiron football to become sports fans themselves.

Think back to your childhood and how much you were impressed with things you saw, heard and experienced from an early age. This is what shapes the personality of the child. A child would likely not have the money or parental permission to just wander off to an NFL stadium 75 miles away, but they might see it on TV flipping through channels and fall in love with the sport, the energy, emotion and teamwork. Without recruiting children at an early age, no sport, hobby or culture stands a chance of surviving throughout the years. So what do you think was the end result of the NFL blackout rule?

Smaller teams entered a death spiral that meant less fans, less merchandise sold and overall less attention received from the public in general. Attention is a resource too and just as important as money, so without attention, NFL teams could stay afloat because of the older fans, who stepped off the bleachers one by one until the public at large became desensitized to the gridiron football completely. In recent years, the most interesting thing about the NFL was the scandals.

In 2004, Janet Jackson revealed her nipple for about half a second during the Superbowl halftime show. This act stirred enough Christians into taking up a quill and parchment to keep the NFL in the public consciousness for a little bit and then it again faded away. In 2016, NFL players protested during the performance of the National Anthem by kneeling, apparently in response to police brutality and racial discrimination[58]. What's tragic is that nobody really talks about gridiron football itself, but it appears the dwindling viewer and attendance numbers have finally caught the attention of the NFL, who started suspending the blackout rule on a yearly basis since 2014.

It took these crusty businessmen and marketers 41 years to finally figure out they've been choking the life out of a sport by intentionally limiting who can watch it. That's not just a coincidence, as these people are so detached from reality that they make all decisions by looking at spreadsheets and business reports rather than talking to fans or even accepting any feedback from anyone except shareholders.

Corporations tend to become unfeeling, uncaring and too big for their own sake but therein lies the key to your personal freedom.

People who make these kinds of business decisions are typically of advanced age and have a solid business background. They built business empires by relying on numbers, facts, spreadsheets and contracts to conduct matters, and while they may be proficient in those, they usually have not the slightest clue about technology. Take Donald Trump, who reportedly doesn't even have a smartphone or know how to write an email, often just shouting things to tweet at his assistants[59]. That's not to say such people can't be successful, it's just that they've made it without technology, which gave them a sort of a blind spot for digital matters. This is where you can hide.

By visiting a website such as Sport Lemon[60], you'll get access to a variety of sports channels for your perusal. What you'll immediately notice is that the address changes as soon as you visit. This is because the original site got taken down through a

DMCA request, but the owner simply acquired a brand-new address, moved the site there and asked the ISP to redirect the traffic. It takes months until the copyright holders figure out where the new address is, during which time it's business as usual. Perhaps a couple thousand unique visitors a month will visit websites such as Sport Lemon, not nearly enough to attract any kind of negative attention or police investigation.

Websites such as Sport Lemon are virtually indestructible, but they don't really use any kind of military grade encryption or special technology. It's merely an ethical hacker who figured out the same principles you've been given in this book, namely that a nimble and smart hacker can exploit the inertia of people in charge for as long as he wants. Copyright holders would have to go hire people to hunt down these sites and go through the legal proceedings to shut them down for good, which would cost money they're not willing to invest, so pirate streaming sites like Sport Lemon live on.

Since it's a pirate website, there are some safety measures you should take. Most of these websites rely on ads as a source of income; not just a few but loads of ads of all kinds. In particular, there are those ads you have to nix to view the content, which involves clicking on them. Don't visit without an adblocker, which will often hide the ad but leave a tiny red "X" for you to click and close it. Malware is rare but not impossible, so do install an antivirus as well. Again, separate machine, yadda yadda.

Content is generally streamed using Adobe Flash, which all major browsers have started to deprecate and won't play automatically, tasking you with one or two extra clicks to "allow Flash this time only". When Flash does finally become obsolete, pirates will simply find another way, don't you worry. There's no HD resolution with pirated video streams, so you'll have to make do with 480p or lower resolutions. The quality isn't great but it's serviceable.

Audio might be warbled, and the commentator might be in Portuguese, Russian, Czech or any other left-field language, including ads in that language during intermissions. It's actually kind of fascinating to listen to a completely different language each time, though you can also find streams in English no problem. Video controls might get stuck in such a way so you can't maximize the video or change the volume, though there's almost always some way to wrestle them into submission.

Delays and buffering issues are common, and you should just get used to them, reloading the page as needed. You might also get a new tab or window opened when you click somewhere inside the page--those are ads trying to load. Channels are usually sorted in categories but the intricacies of how they work are up to you to discover. One more important aspect of pirate sports streaming websites is finding a community of like-minded individuals.

Plenty of websites related to these sports streaming pirate websites feature a simple bulletin board messaging system and a semi-anonymous real-time chat box where you can put in any kind of nickname and start messaging other people. There is usually no need to give an e-mail address or other such nonsense, so you can create a new username each time you visit and remain anonymous.

Don't share personally identifiable information, such as your name, age or location, since that can be used for doxxing, meaning discovering where you live to do damage to you. If you must share some personal information, lie all the time until the people prove themselves trustworthy. However, if you do need help with the related streams, do feel free to ask for advice.

By finding a small community of online enthusiasts that share your passion for ethical hacking, usually having 20-100 people, you can create much more intimate relationships than if herded in Facebook's paddock with several thousand accounts that are your "friends". You'll encounter moms, dads, grannies or just some kids passing time in a chat box, so keep an open mind and don't presume anything. You'll realize that there's an entire ecosystem of these websites, chats and communities that are surprisingly welcoming and tolerant. Huddling around a shared hobby, now that's the recipe for creating a wholesome community you'll want to engage with.

Another great sports streaming website is Feed2All[61]. It works in an almost identical manner to Sport Lemon. As the title banner implies, the former name was First Row Sports and the "Notices" link at the bottom of the page lets copyright holders claim any given video stream as their own. So, will Feed2All comply with DMCA takedown requests? Just like we discussed in the DMCA section, the safe harbor provision was left in to allow the big boys like Google to provide services. Without it, Google wouldn't be able to display those couple million results if even one of them contained a copyright infringement, but it requires responding immediately and without question to copyright claims.

Feed2All's owner simply figured out that allowing DMCA takedowns will make his website eligible for the safe harbor provision, thus making it a thousand times more impervious to DMCA. If copyright holders want to sift through the ad-infested streams and issue takedown requests all day long, sure, more power to them. Another interesting tidbit on the same page is the disclaimer "videos are not on our servers." It turns out that content aggregators, such as Sport Lemon and Feed2All are immune to DMCA[62] due to not actually hosting any content. As long as whoever owns the content states, "Yeah this stream is mine," and nobody challenges it, the DMCA does not apply.

The owner of Feed2All isn't as much a pirate as a business genius. This terrifying DMCA act that makes tech giants tremble on their thrones turns out to be an easily avoidable obstacle with just a single link and two sentences as disclaimer. Now, the owner of Feed2All might get hoisted by his own petard if copyright holders send an avalanche of DMCA claims his way, since he's probably a one-man team and he'd probably get the police kicking down his doors if Feed2All ever became popular. That's not an exaggeration, as other people have tried dodging DMCA using the same

content aggregator loophole and got badly
burnt.

Chapter 13: Mega Upload And Anonymous

Kim Dotcom is the chosen alias of one Kim Schmitz, a chubby ethical hacker who just wanted to make the world a better place by creating a website where users could upload any file and share the link with their friends. You can probably tell where this is going – people started uploading copyrighted material to his website and he got shut down through DMCA. That's almost exactly what happened but not quite, as truth is much more incredible than fiction.

In 2005, Kim Dotcom created a file hosting website called MegaUpload – free users had to wait for downloads and had throttled download speed, but it worked just fine, attracting all sorts of pirates. It ran for 7 years until getting shut down, making money by selling premium account features to users and getting about 50 million daily visitors. Originally of German and Finnish origin, Kim moved to New Zealand where he lived like a king on a compound with a mansion that had a panic room. Why New Zealand? Well, it was one of rare countries that didn't enforce copyright. In 2012, Auckland police, under the auspices of FBI, raided[63] Kim's estate, seizing a couple shotguns needed to later justify the use of force, art, cash and assets worth about $16mm.

Video of the raid[64], recorded 6:40 am, 20th January 2012, shows how an Auckland police helicopter approaches and lands in front of Kim's mansion, unloads four policemen and takes off, recording the action below. The estate gates are open as Kim was waiting for guests and didn't find it strange a helicopter landed. Immediately after the helicopter takes off, several police vehicles stream through the gates and a total of 76 policemen start combing the estate. Once Kim hears commotion at the entrance, he quickly hides in the panic room but doesn't lock the doors because he knows he'll probably get shot when they do break inside.

The police takes 13 minutes to locate the panic room. Kim's pregnant wife, three children, guests, staff and security are arrested. Kim is thrown to the ground with boots and knees on his hands and back for pain compliance as police dogs scour the compound. Charges involve racketeering, copyright infringement and money laundering, with possible jail time for Kim and other Megaupload executives for up to 50 years. The most important question is: why was FBI involved with an arrest of a German national in New Zealand? It's about a show of force to terrorize those who might want to do the same as Kim and sending a message – we will get you no matter if your country doesn't care about copyright.

Auckland police would have probably just invited Kim down to the station for a cuppa tea and a chat, but FBI ensured that unnecessary force was used for the whole world to see that copyright, this arbitrary notion of owning ideas and concepts, is untouchable. You do not make money off of anything that benefits US companies, in particular the Recording Industry Association of America (RIAA). Favors were asked, backs were scratched, and the raid went as it did.

Warrants were so broad that the police could simply pack up everything that wasn't nailed down, especially electronic equipment. Later court investigation will reveal that FBI copied all hard drives from Kim's mansion without authorization and took the copies with them to the US, probably at the request of their friends at RIAA. Kim spent the first night in jail with hardened criminals.

After a series of legal proceedings that culminated with his case ending up in front of New Zealand Supreme Court, that did rule the raids legal, the following were determined:

• the search warrants were too broad

• giving the hard drives to the FBI was in breach of protocol

• Kim was spied on by the New Zealand intelligence agency, for which he got an apology from the Prime Minister

Kim would later settle with the Auckland police but suing the spy agency that facilitated his arrest turned out to be much trickier. Being a prolific internet user, Kim's Twitter feed[65] holds all the latest and most relevant happenings in his life, as well as a wealth of anti-US government rants, to which Kim is fully entitled because, as he puts it, "I never lived there, I never traveled there, I had no company there but all I worked for now belongs to the US."[66] This was done under asset seizure regulation that pretty much allows the US authorities to take whatever assets they wish with no recourse as long as there's some kind of implied crime involved.

Anonymous, the shadowy group of ethical hackers, retaliated promptly to Kim's arrest and took down the US Department of Justice website as well as RIAA's and others. That wasn't the first nor the last time Anonymous performed such retaliatory attacks, but it's hard to attribute actions to a group anyone can belong to. Some persons even organized real-life protests with Guy Fawkes, the protagonist from the 2005 V for Vendetta movie, masks but there's no telling if they were affiliated in any way with actual hackers or were just bored teens.

In 2008, Anonymous performed what's known as Project Chanology, which was an attack on Scientology coordinated on the popular imageboard 4chan, a chaotic forum rife with gore, nudity, porn, profanity and vivid fantasies of all kinds. No registration is required and though it's possible to create a unique username and attach a hash to it, such practice is normally discouraged. It's hard to tell if anyone is actually posting real information on 4chan, so best presume it's all just a LARP, or live-action role play, unless proven otherwise.

No tutorial or operating manual exists for using 4chan, and each user is meant to learn on his own by observing and taking in the atmosphere to blend in or as 4chan users would put it, "Lurk more". Another common trait of all 4chan users is that they refer to 4chan as having a different purpose every other time, such as "Mongolian fly-fishing forum" or "Taiwanese pottery forum". All of this is meant to instill a sense of anonymity and humility in fresh users.

It sounds impossible, but 4chan boasts a steady rate of production of new content that eventually goes viral and affects the public consciousness. One urban legend states that it was 4chan that first realized

Donald Trump would become the president, mashing all sorts of memes to make Trump appeal to young, disenfranchised voters. There is something to 4chan that makes it worth visiting, some genuinely exhilarating quality that only happens when smart people get together and discuss whichever thought crosses their minds.

The site itself is organized into boards, each dedicated to a different topic, such as books, GIFs, weapons or fitness. Moderators do exist on 4chan, as unlikely as that sounds; they're mostly unpaid volunteers mockingly called "janitors". The most important quality of 4chan is anonymity, which is why users call each other "anon" rather than by any other moniker. Nobody cares about who you are but only what you have to say and show. Ideas are mercilessly mocked or heartily praised based on their merit. It's in this kind of environment that Anonymous emerged and started to attack Scientology.

In January 2008, a Scientology video featuring Tom Cruise found its way to YouTube only to be quickly claimed as copyrighted material by Scientologists. This infuriated Anonymous, who organized denial-of-services attacks on Scientology websites, prank faxes to waste their ink and prank calls to take up their phone lines, hurting their recruitment drives. All of these attacks were in line with what we learned so far about hacking–cheap, simple and low-key. Project Scientology culminated with protests all over the world meant to increase awareness of harsh recommendations the Church gives to new members namely to cut out all family and friends. The Church is known to viciously pursue and harass anyone who dares criticize it, hence why Anonymous wear masks during protests.

In March 2008, someone posted a series of flashing videos on a forum for epilepsy support, triggering seizures in visitors. There was supposed evidence that Anonymous performed the attack, with circumstantial evidence claiming 4chan was used as a staging ground. 4chan janitors claimed the attack was actually organized by the Church to smear Anonymous, though 4chan might have been used in the process.

In April 2011, Anonymous attacked Sony for their aggressive anti-piracy stance, taking PlayStation servers offline, which made it impossible for users to access their digital game libraries. Sony employees had their e-mail accounts hacked and were doxxed because the company sued a hacker that made a tool that allowed PlayStation 3 users to run pirated software. Also, in 2011, someone posing as Anonymous warned Westboro Baptist Church they'd be next in the line, to which WBC responded with insults and hate speech, as is common for them.

In 2017, Anonymous hacked into Tor servers and presented a message to visitors to about 10,000 Tor-only websites bragging about it. Those servers had about 20% of all Dark Web, meaning sites that were not reachable through mainstream web browsers, but Anonymous claimed nearly half of all data they found on the servers was child porn.

It's hard to know where Anonymous begins and ends, though it's likely there have been numerous hacks done in their name just to smear them and justify prosecution. Beware of joining "ethical hacking" organizations, as idealistic as they might seem, since they're likely to be crawling with agent provocateurs. Their job is to infiltrate any organizations that might even conceivably conspire against the government and provoke them into taking violent action that is guaranteed to fail because the police will appear at just the right moment.

An agent provocateur can be recognized by:

• having a large bankroll while being unable to explain how he got it or what he does for a living (he's financed by the government)

• no friends, social connections or habitable quarters (that's his front)

• always being the loudest in the group and suggesting violent action (that's how he brings down the group)

• trying to set himself as a leader when the group never had or wanted one

• having skills that no ordinary person has, such as knowing how to make explosives (this is how he radicalizes a group)

- always trying to collect information about members of the group (to find out incriminating evidence)

Admittedly, this checklist applies more to an agent provocateur infiltrating real-life organizations, but you never know who you're dealing with unless you pay attention to the signals. This implies that the government is the one afraid of people, not the other way around. The only reason why gramps has to be groped at the airport by the TSA is that the government is so paranoid that he's considered a potential insurgent, just like the rest of us.

All the surveillance, the unfettered wiretapping and collection of metadata make sense now, don't they? Just like the old businessmen running corporations, those inside the government are clueless about reality and have no idea what people think, but they fiercely strike back if someone hurts their pride. Oh, do they.

Chapter 14: Julian Assange

Here's a question for you – what's the punishment for having sex with a woman and the condom slipping off or breaking? That's a trick question, because that's what happened to Julian Assange. He ended up wanted by Swedish authorities, trapped inside an Ecuadorean embassy in London for years on end, with Hillary Clinton supposedly threatening to drone strike him. It's not that she is such a champion for women's reproductive rights but rather that Julian happened to be at the head of an organization that caused immense grief and embarrassment to the US, which waited for any excuse to make him pay.

One breezy day in Sweden, Julian was giving a speech at an ethical hacking conference. Trying to stay low, he let like-minded individuals drive him around and house him. One of the young gals at the conference fell for the silver fox's charm and he decided to spend a night at her place. The aforementioned condom incident happened but Julian just kept going and she apparently didn't protest or thought it a big deal. Next morning, she contacted a feminist friend and asked if what happened was OK, to which the friend replied that it's a big deal and she should report him to the police. "You just got shafted by the patriarchy!" she was told.

They both went and started making a report, but the first girl got cold feet and no longer wanted to file it. This had gone way too far and now she had to commit all the saucy details to paper. The word had already gotten to the Swedish chief prosecutor that Julian Assange was involved in what was essentially a sex crime, so she issued a warrant on his name. At that moment, Julian was giving a speech in London when he heard that Sweden wanted him for some crime. He immediately thought it was about Wikileaks and, rightly fearing Sweden would extradite him to the US, sought asylum at an Ecuadorian embassy, where he stayed for years while fighting in court.

If Ernest Hemingway and Oscar Wilde were to join forces from beyond and write a story, it wouldn't be this unbelievable. Despite being on a literal island, which he can't leave unless he grows wings, there's been a number of police patrols stationed outside the embassy 24/7, costing the UK taxpayers millions of pounds in surveilling a single man who hasn't committed any crime. In Sweden, the condom slippage is at worst fined the equivalent of a couple hundred dollars. Cooped up inside the embassy, Julian contacts the outside world via Skype, schedules a few interviews and even gets Pamela Anderson to visit him. So, what was the big deal with Wikileaks?

Standard procedure for insiders in the US government who notice breach of protocol is to report it to their superiors. If the protocol is being breached by those same superiors, there's nothing a whistleblower can do as revealing classified information puts the person that knows it in danger as well. Mainstream media won't lift a finger to help such a whistleblower, instead working with the authorities to expose the whistleblower and throw him in jail.

Civilians don't quite get how it works – everything the government does is classified information, with tidbits being revealed only on a need-to-know basis. This neatly proves that all the outrageous news about Donald Trump pouring from the mainstream media is completely fake, because any actual information about anything regarding him as President would be classified and those publishing it would be brought to justice for doing so.

How many pounds of rice Donald Trump eats every day or if he eats any rice or if he eats anything at all is classified information that can get the leaker in serious trouble. Why? Well, if you knew Donald Trump loves eating rice, you might conceivably try to poison the entire US supply in some manner to hurt him. Makes sense why these irrelevant tidbits are kept secret. Law enforcement, military and intelligence operations are all shrouded in a bulletproof veil of secrecy and even the fact the shroud exists is classified. Whenever you hear a conspiracy theory, just ask yourself, "Would this be classified information?" If yes, then it's likely completely false, although nothing in life is binary except computer code.

Anyway, people at the top of such organizations, those who can connect the dots and parcel out information to cover their tracks, can do whatever they want and nobody can touch them. Julian Assange founded Wikileaks to root out corruption, wickedness and crony collusion from all government bodies by giving whistleblowers a way to leak information that the general public must know. By using verification procedures Julian never reveals to anyone, he can make sure only those with actual classified information get to post to Wikileaks to drain the swamp, as it were.

In 2010, Wikileaks published a 2007 footage that shows US helicopters mowing down Iraqis and laughing about it. Detailed expense and casualty logs from Afghanistan were leaked too along with the Iraq ones, allowing the world to see some 15,000 casualties that were classified, probably because they were involved in unsavory classified operations. That same year, Wikileaks revealed secret diplomatic cables between US embassies and homeland officials and over 700 files on Guantanamo prisoners in 2011, showing "enhanced interrogation" techniques used, such as waterboarding that simulates drowning.

This so riled up the US authorities that they decided to start investigating Assange under claims of espionage. Later the charges would balloon to wire fraud, because why not, general conspiracy, theft of government property and so on. So, Assange fought Swedish charges until 2012, when all his appeals were exhausted. He applied for asylum in 2012, waiting for the statute of limitations on his Swedish charges to expire, which is due to happen in 2020. After that, he should hopefully be able to just walk out and leave London like a free man. What's sad is that the Australian government had de facto abandoned Julian, publishing a letter that already proclaimed him a criminal and stating he could always ask to be transferred to an Australian prison.

Wikileaks was leaking at full power while Assange was fighting for his freedom. Edward Snowden revealed his trove of information on Wikileaks, which showed how everyone, including foreign heads of state, was being spied on routinely using NSA's infrastructure. Five million e-mails from Stratfor, surveillance documents, secret cables, transnational agreements, intercepts and plenty more just kept coming. We mentioned how harsh the rules on classification are, but with these leaks, it was like there was someone inside the NSA leaking it. One possibility is that insiders, dismayed by the corruption and lack of transparency, decided to put their feet down and use the skills and training they already had to expose their dirty overlords.

In 2016, Wikileaks published a set of leaked e-mails from the highest echelon of US officials, including John Podesta and Hillary Clinton. These revealed some unusual behaviors, such as Podesta's love of pizza, but also helped show collusion and lackadaisical behavior when it came to handling classified information. Analysts will later claim these leaks will in large part help elect Donald Trump. Among the first promises he made when entering office, Donald Trump said he'd fight the deep state and drain the swamp, referring to the corrupt officials and indirectly thanking the honest people, the ethical hackers working for the US government that at times risked their lives and livelihoods to bring forth the truth the whole world should know about.

Chapter 15: Patents

When copyright is strictly applied to inventions, it's called a patent. The same restrictions as copyright apply, except that patents are even more nebulous, fostering legal disputes that companies love. A competitor that's involved in a patent lawsuit is slowly being bled dry. A tech giant can afford to buy up a whole array of patents applying to any given field to create what's known as patent minefield, a dangerous place for curious people. Anyone trying to create a new device can be pestered by the company's legal team or, if they carry on, sued for damages and potential acquisition of their product. Is there anything better than getting a product for free? One such company that specializes in suing people over patents is Unwired Planet, holding over 2,000 patents from Ericsson, who handed them over in return for a piece of any money they can get.

They were an actual company at one point, hiring and firing, until they realized there's much more money to be made in suing people. In 2015, Unwired Planet sued Samsung, Google and Huawei over their smartphones[67], claiming they violated some of those 2,000- plus patents and asked for millions of dollars in royalties. In 2013, they tried to do the same to Apple but lost in court. In 2014, they actually got $100mm from Lenovo in a licensing deal and also launched lawsuits in the UK and Germany to probe for weakness. Still, Unwired Planet is little league for Intellectual Ventures, a US-based company that nearly invented the term patent troll, meaning someone who legally bullies others over patents to extract money. When Samsung, Google, Huawei and Apple can't be intimidated, it's time to sue some florists.

Intellectual Ventures sued a florist in 2016 over using a dispatcher system, claiming their patent portfolio contains exactly that idea. Electronic Frontier Foundation, a non-profit that fights for cyber-freedoms, slammed the case as "stupid patent of the month"[68]. This kind of patent is what's commonly known as "on a computer", where you take a simple action and add "on a computer". For example, sorting images on a computer and arranging folders on a computer thus become two separate patents, albeit weak but still sounding technical enough to scare defendants into paying. The Supreme Court ruled this kind of patent invalid[69] in another patent troll case that dragged on through courts for 6 years, stating that "merely requiring generic computer implementation fails to transform that abstract idea into a patent-eligible invention". Still, money could be made by getting sued by these patent trolls.

The concept of vexatious litigation states that if a defendant constantly files lawsuits to annoy, harass or threaten legal costs, all of which is what these patent trolls are doing, and he keeps constantly losing, the defendant can sue back for damages. Intellectual Ventures has likely filed thousands of lawsuits and barely looked at who they're suing, so if someone were to make fake business ventures that baited them into suing, it's likely this would qualify as vexatious litigation. Of course, that would be such a terrible mockery of the US legal system but such a fitting punishment in line with ethical hacking.

Chapter 16: Penetration Testing

Penetration testing is a white hat technique to assess a company's vulnerability to hacking. White hats can be commissioned by managers or senior executives to roleplay as new hires and given a few weeks to try and hack into systems. But when a company wants to look good, it can tip off the chief of security and the entire thing looks like the Kobayashi Maru test from Star Trek–a test where the taker is designed to fail.

One white hat codenamed Tinker was hired in November 2018 to do a penetration test for an unnamed company[70]. The goal was to assess how much damage a malicious employee with some know-how could do. Tinker was given a week to hack into whatever he could, so he quickly completed the onboarding procedure and brought in his own Linux laptop with hacking tools. This let him scan devices, the network, and all the accounts trying to figure out how to get in, but only a couple of most trusted IT staff had accounts authorized to see anything or make any changes.

From one failed attempt to another, our white hat sweated bullets as he dodged the piercing gaze of IT staff and pretended he was doing useful work. His tools yielded nothing, causing him to spend hours trying to break through passwords or reveal any vulnerabilities, but to no avail. Finally, as if in some surreal theater play, he was revealed to everyone in the office by an old lady who saw him "typing at the computer".

Real penetration might go something like this – a forgotten account belonging to a field technician is discovered on a network and used to create a working clone of a tech support voicemail where hackers respond to tech problems of people on the road to ultimately get their usernames and passwords[71]. When one such company employee inquires about a weird VPN problem, the hacker realizes he knows the solution and helps out. The employee is thankful and would never doubt or question the guy who just helped him out – why, he's on our side! Just like in any other system, the weakest link is people.

Social engineering refers to exploiting the inherent trust all people show to those in positions of authority. This trust stems from the urge to help others who have status, thus expecting to gain status and get helped in return at some point in the future. By doing small favors and positioning himself as an authority, a hacker can gain access to usernames, passwords, files, networks and accounts with no effort. It helps if the hacker is seen as part of the team, which exploits another human weakness known as "in-group preference," where we see those belonging to our inner circle as more trustworthy than others. Now let's compare the two cases of penetration testing to see what made one fail and the other succeed.

In the former case, Tinker was put in a position of distrust from the start and did nothing to elevate his position or socially engineer others around him. It could be that an old lady in the accounting department had access to a network or an account that could be used to hurt the company, but Tinker did nothing of the sort, obsessing over his tools and passwords. He could have been given ten years to break through, but it wouldn't have mattered because nobody trusted him in the first place and any mishaps, even ones he didn't cause, would have been blamed on him. Obviously, black hat hacking is a crime, so the entire point is to get away with it, which means people should never suspect you.

In the latter case of penetration testing, the trick was in the passive approach of creating a trusted voicemail number associated with the company and waiting for users with a technical problem. Such users are frustrated, anxious and upset, so when the hacker helps them out in any way, they'll feel gratitude and reveal personal information they shouldn't. While Tinker lost his sanity trying to break through, successful hackers take their time and slowly gain ground, in this case a forgotten user account that also means there was no risk of getting caught, which is crucial for hackers of all kinds.

Another brilliant successful penetration test had to do with security cameras installed inside the company offices. White hats realized cameras were accessible over the internet, so they examined them until finding a zero-day weakness, meaning an unpatched vulnerability unknown to the camera maker. They hooked into the live feed and found out they could look over the shoulder of every employee in the offices as they're typing in their usernames and password. People ignore cameras anyway, so hackers could take their time and collect as much information as they wanted, including confidential info. Again, without arousing suspicion, raising any alarms and with minimal technical prowess, our white hats gained a passive advantage and slowly exploited it.

The most devastating hacks are done through positions of authority everyone trusts and using hardware everyone ignores. Once we realize how hackable everything is and that there are people all over the place researching, hacking into and observing what we're doing, we'll change our demeanor. The solution is to distrust everything, including official support staff and constantly conceal what you're doing, even from cameras your boss installed in the offices. Security and accessibility are inversely correlated, so you'll either have one or the other but not both. The trick is in finding such security that is accessible to you more than to anyone else.

Internet is another problem, as it makes remote access much easier. It's acceptable to have internet access if you're actively using it, but hardware with such capability that doesn't fulfill any productive purpose is a huge problem. So, if there's anything where you don't really need the internet, turn it off.

Chapter 17: Types Of Hackers

A hacker can be placed into one of the following categories: black hat, grey hat or white hat. Each hacker is classified based on their intent. These terms are borrowed from the Old West when a good cowboy would wear a white hat while a bad cowboy would don a black hat.

White Hat Hackers

A white hat hacker, who is also called an ethical hacker, does not want to harm the system. His motive is to identify the weakness in any network system or computer through different vulnerability assessments and penetration testing. Ethical hacking is legal, and, in fact, many companies hire ethical hackers to find vulnerabilities.

Black Hat Hackers

A black hat hacker, who is also known as a cracker, is someone who wants to hack a network or a system to gain unauthorized access. This type of hacker wishes to harm the system or steal some sensitive information. Black hat hacking is illegal since the person who is hacking the system does it with bad intentions. This includes violating privacy, blocking any communication on the network, stealing corporate data, damaging systems, etc.

Grey Hat Hackers

A grey hat hacker is a blend of both a white hat and a black hat hacker. These hackers do not have any malicious intent but hack a network or a system merely for fun. They want to exploit the vulnerabilities in the system without actually taking permission from the owner. Usually, their goal is to inform the owner of any weaknesses and gain appreciation and/or a sum of money from them.

Miscellaneous Hackers

Apart from the list of hackers detailed above, there are a few other categories of hackers that should be mentioned. These include script kiddies, intermediate hackers, elite hackers, hacktivists, cyberterrorists, and hackers involved in organized crime.

Script Kiddies

These hackers are computer novices who use the different tools and documentation available on the Internet to perform a hack. They do not know what happens behind the scenes and only comprehend enough to cause minimal harm. They are often sloppy, so they leave digital fingerprints everywhere. These are the hackers you hear about in the news. They need very minimal skills to attack a system since they use what is already made available to them.

Intermediate Hackers

These hackers know just enough to cause some serious issues. They have knowledge about networks and computers and use this knowledge to carry out well-known exploits. Some intermediate hackers want to be experts at the process; if they put in some effort, they can certainly become elite hackers.

Elite Hackers

Elite hackers are experts. They're the people who develop several hacking tools and write scripts and programs. Script kiddies use these very tools and programs to perform their own attacks. Elite hackers write codes to develop malware like worms and viruses. They know how to break into a system and cover their tracks or pretend that someone else was responsible for the attack.

Elite hackers are secretive and only share information if they believe that their subordinates are worthy. For some lower-level hackers to be evaluated as worthy, they should possess some special information that an elite hacker can use to perform an attack on a high-profile system. Elite hackers are the worst type of hackers. However, there are not too many of them in the world when compared to the number of script kiddies.

Hacktivists

These hackers disseminate social or political messages through their attacks. A hacktivist always finds a way to raise awareness about a given issue. Some examples of hacktivism are the many websites that had the "Free Kevin" messages. These hacktivists wanted the government to release hacker Kevin Mitnick from prison. Some other cases include the protests against the U.S. Navy Spy Plane that collided with a Chinese fighter jet in 2001, attacks against the U.S. White House website for years, hacker attacks between Pakistan and India and messages supporting the legalization of marijuana.

Cyberterrorists

Cyberterrorists attack government computers or other public utility infrastructures like air-traffic control towers and power grids. They steal classified government information or crash some critical systems. Countries have started to take cyberterrorist threats seriously, ensuring that power companies and other similar industries always have information-security controls in place.

These controls will protect systems from such attacks.

Organized Crime

Some groups of hackers can be hired to perform an organized crime. In 2003, the Korean police busted one of the largest hacking rings on the Internet. This ring had close to 4,400 members. In addition to that group, the Philippine police busted a multimillion-dollar hacking ring that sold cheap phone calls made through the lines that the ring had hacked into. These types of hackers are always hired for a large amount of money.

Ethical Hacking Terminologies

This chapter briefly details some of the common and important terms that are used in the field of hacking.

Adware

Hackers use this software to display advertisements on a system by force.

Attack

Hackers perform this action to access a system and extract some sensitive data from that system.

Back Door

A back door, which is also referred to as a trap door, is an entry port into software or a computer. This port does not require any login information or a password, and as a result, it can bypass all security measures.

Bot

A bot is a type of program that is used to automate any action, thereby increasing the number of times it can be performed. This means that the bot will perform the function for a longer time when compared to a human operator. For instance, hackers use bots to call a script that can be used to create an object or send an FTP< Telnet or HTTP file at a higher rate.

Botnet

Botnets, which are also called zombie armies, are a group of computers that can be controlled without the knowledge of the owner. These are used to perform denial-of-service attacks or send spam.

Brute Force Attack

A brute force attack is possibly the simplest attack that a hacker can perform to gain access to a system or application. This attack is an automated attack, and this means that it will try different usernames and passwords repeatedly until it can access the system or application.

Buffer Overflow

The buffer overflow is a flaw that can be observed when a lot of data is written onto a single block of memory. This means that the memory can no longer hold onto that data.

Clone Phishing

Clone phishing is a type of legitimate and existing email that has a false link. This link will trick a recipient into providing some personal information that the hacker can use to disarm the system or network.

Cracker

A cracker is a type of hacker that modifies any software to access some features of a system, such as copy protection features.

DoS or Denial-of-Service Attack

A denial-of-service, or DoS, attack is used by a hacker to ensure that a network resource or server is not available to the user. This is done by suspending the services of that server or resource.

DDoS

DDos stands for distributed denial-of-service attack.

Exploit Kit

An exploit kit is a system that a hacker designs to run on some web servers. This system is used to identify any vulnerabilities in a client machine that is communicating with the web server. It will then exploit those vulnerabilities and afterwards, execute some malicious code in the system.

Exploit

An exploit is a part of code or a chunk of data or software that will take advantage of a vulnerability or a bug in the system and network which, in turn, compromises the security of that system or network.

Firewall

A firewall is a type of filter that is placed on a network. This filter helps to keep unwanted intruders away from the system or network. In addition, it will ensure that the communication between the users and systems inside the firewall are safe.

Keystroke Logging

Keystroke logging is a process during which a hacker tracks how the keys are pressed on the keypad. This process will help the hacker develop a blueprint of the human interface. It is often used by both black and grey hat hackers to record some passwords. A keylogger is most commonly delivered onto a system using a phishing email or a Trojan horse.

Logic Bomb

A logic bomb is a type of virus that is added to a system that will trigger a malicious attack if some conditions are met. A common example of a logic bomb virus is a time bomb.

Malware

Malware is a term that describes a variety of intrusive and hostile software, including Trojan horses, spyware, scareware, adware, virus, ransomware, worms and any other malicious programs.

Master Program

Master programs are those programs that black hat hackers use to transmit commands into zombie drones (explained later in the chapter). These drones carry spam attacks or denial-of-service attacks.

Phishing

Phishing is a fraud method where the hacker sends an email out to the target. The hacker will use this email to gather some personal or financial information from the user.

Phreaker

A phreaker is a normal computer hacker. These hackers often break into telephone networks and either tap the phone lines or make long-distance phone calls.

Rootkit

Rootkit is a software that is often malicious. A hacker designs this software to hide some processes or programs from any normal detection method. This will ensure that the rootkit is stored on a system and has privileged access to the system.

Shrink Wrap Code

A shrink wrap code attack is a way to exploit the holes in a poorly configured or unpatched software.

Social Engineering

A hacker uses social engineering to deceive another person. The hacker uses this technique to acquire some personal information about the user, like credit card details or passwords.

Spam

Spam is an unsolicited email. This is also called junk email and is often sent to a large group of people without their consent.

Spoofing

Spoofing is a technique that a hacker uses to gain access to a system or network. The hacker will send a message to the computer using an IP address, and this address will indicate to the system that the message is being sent from a trusted host.

Spyware

Spyware is a software that's used to gather information about an organization or person without their knowledge. This software can be utilized to send sensitive information to any entity without the consent of the customer. It can also be used to assert control over a system.

SQL Injection

SQL injection is an injection technique code that is written in SQL. This tool is used to attack any data-driven application. It includes some malicious SQL statements that are entered into a field in the data. An example of an SQL injection would be dumping all data into the attacker's folders.

Threat

Threats are possible dangers to a system or network. These can be used by hackers to exploit a vulnerability and compromise the security of a network or system.

Trojan

A Trojan horse, or Trojan, is a program that is designed to look like a normal program. Differentiating between a Trojan and a regular program is difficult. This tool can be used to alter information, destroy files and steal sensitive information like passwords.

Virus

A virus is a piece of code or a full program that is malicious. It copies itself in the system and has a detrimental effect on it as a result. A virus can both destroy data and corrupt the system.

Vulnerability

A vulnerability is a weakness in the system or network that allows a hacker to compromise the security of that system or network.

Worms

Worms are like every other virus in the sense that it can replicate itself in the system. It only resides in the active memory but does not make any changes to the files and will only duplicate itself.

Cross-Site Scripting

Cross-site scripting, or XSS, is a security vulnerability that is often found in a web application. This vulnerability gives the hacker the ability to inject some script into a web page that is viewed by users.

Zombie Drone

A zombie drone is used by hackers as a soldier to perform a malicious activity. This drone is a hijacked computer that is used by some hackers to distribute unwanted spam emails.

Ethical Hacking Tools

Now that you know what ethical hacking is, let's look at some of the different tools that are available for you to use to prevent any unauthorized access to a network system or computer.

Nmap

Nmap, or Network Mapper, is a tool that is used for security auditing and network discovery. It is an open source tool that was designed to scan a large network. It also works well with single hosts. A network administrator is used for different tasks, including managing service upgrade schedules and network inventory and monitoring service or host uptime.

Nmap can determine the following using raw IP packets:

•   The different hosts available on the network

•   The operating systems that the hosts run on

•   The different services offered by those hosts

•   The different firewalls that the hosts use and any other characteristics

This tool can run on most operating systems, including Linux, Windows and Mac OS X.

Metasploit

Metasploit, another powerful exploitation tool, is a Rapid7 product. Many of the resources used can be found on the source website: www.metasploit.com. The tool has a commercial and free version and can be used with Web UI or command prompt.

You can perform the following operations using Metasploit:

•   Penetration tests on small networks

•   Check the vulnerability in some systems

•   Discover any import or network scan data

•   Run individual tests on a host or look at the different modules that one can exploit

Burp Suite

Burp Suite is a tool that's used by both malicious and ethical hackers to perform a security test of any web application. This suite has different tools that work together to support the process of testing, right from the mapping to the analysis of the application's surface. It's often used to exploit or locate any vulnerabilities in the application. This suite is simple to use and gives an administrator full control to combine different techniques to improve testing. Burp can be configured easily, and it has different features that can help an experienced tester with their work.

Angry IP Scanner

Angry IP scanner is a cross-platform and lightweight port and IP address scanner. This tool can scan an IP address in any range and can be used or copied anywhere. It utilizes a multithreading approach to increase the speed of scanning. In this approach, a separate scanning thread is employed for every address. Angry IP scanner checks if an IP address is active by pinging the address, and it will then determine the MAC address and scan ports and resolve the hostname. The data that is gathered using this tool can be saved to an XML, TXT, IP-Port List or CSV file. You can gather information about any IP using this tool.

Cain and Abel

Cain and Abel is a tool used in Microsoft Operating Systems for password recovery. This tool helps to retrieve passwords using one of the following methods:

• Recording a VoIP conversation

• Sniffing the network

• Decoding a scrambled password

• Cracking an encrypted password using Brute-Force, Cryptanalysis and Dictionary

• Revealing a password box

• Recovering wireless network keys

• Uncovering a cached password

• Analyzing routing protocols

This is a tool that most professional penetration testers and security consultants use for ethical hacking.

Ettercap

Ettercap, or Ethernet capture, is a network security tool that's used for a man-in-the-middle attack. This tool can sniff live connections, filter any content on the fly and perform other interesting activities. Ettercap has numerous features that can be used for host and network analysis and supports the dissection of protocols (both active and passive). It runs on many operating systems, including Mac OS X, Linux and Windows.

EtherPeek

EtherPeek is a tool that helps to simplify network analysis that is performed on a heterogeneous network environment. This is a very small tool that can be installed on any system in a few minutes. One can use this tool to sniff the traffic packets on any network and supports different protocols, including IP, AppleTalk, UDP, NBT packets,

IP Address Resolution Protocol (ARP), NetBEUI, TCP and NetWare.

SuperScan

SuperScan is a powerful tool that can be used to resolve hostnames and scan any TCP ports. It has a user-friendly interface that can be used to perform the following functions:

•   Port or ping scan using a different IP range

•   Scan different ports in the network using a built-in or random range

•   Decipher the responses from different hosts connected to the network

•   Modify the port description and list using a built-in editor

•   Merge different lists to build a new one

•   Connect different open ports

•   Assign a helper application to a port

QualysGuard

QualysGuard is a suite of tools that can be used to lower the cost of compliance and simplify any security operations. This tool can automate the full area of compliance, auditing and protection for web applications and IT systems. QualysGuard can deliver some critical security intelligence and includes a variety of tools that can be used to detect, monitor and protect the network.

WebInspect

WebInspect is a tool used to assess an application's security. This helps to identify any unknown and known vulnerabilities that exist in the application layer for any tool. It can also be used to check if a server has been configured correctly and helps to test the vulnerability of a system using attacks like cross-site scripting, parameter injection, directory traversal and others.

LC4

LC4 (formerly called L0phtCrack) is a password recovery and auditing application. This tool is used to test the strength of passwords and to sometimes recover a password on Microsoft Windows by using hybrid, brute-force and dictionary attacks. LC4 is used to retrieve lost Windows passwords, which will help to streamline the process of migration. It also assists in retrieving a lost password for an account.

LANguard Network Security Scanner

A LANguard network security scanner scans a network to identify the devices connected to it. It also provides some information about every node in the network. Using the LANguard network scanner, one can obtain any information about the operating system that's used by every system connected to the network. This tool is also utilized to detect any registry issues and can provide a report in HTML format. You can obtain information regarding the NetBIOS name table, the MAC address and the user logged into the network using this tool.

Network Stumbler

Network Stumbler is a WiFi monitor and scanner that is used on the Windows Operating System. This tool allows a network professional to detect a wide area network. Most hackers utilize this tool to find a wireless network that is not used for broadcasting. Network Stumbler can help you verify if a network has been configured well, detect any interference between wireless networks and test the signal coverage and strength. Additionally, it can be used on any unauthorized connections.

ToneLOC

ToneLOC, or Tone Locator, is a program that was written in the early 90s for MS-DOS. It was used in war dialing computer programs. Through war dialing, one can scan phone numbers using a modem and dial every number that has the same area code. Malicious hackers use this tool to breach security by identifying modems that can be used to enter a network or computer system or guess a user's account. Ethical hackers can use it to detect any unauthorized device on the computer's network.

Chapter 18: Ethical Hacking Skills

This chapter covers the ten most important skills every hacker needs to possess and consistently improve on to become a professional in the field.

Basic Computer Skills

You are probably laughing at this skill; however, it is extremely important for a hacker to understand the basic functions of a computer. You'll need to learn how to use command lines in windows and also understand how to edit the registry and set the networking parameters. These may seem like simple skills, but they're actually very difficult to master. If you make an error in the command line, you will mess up the entire hacking process and make the system more vulnerable than it initially was.

This is a skill that professional hackers build on every chance they get. They believe that there is always room for improvement. Amateurs, on the other hand, may believe they have learned everything there is to about computers and rarely build on the knowledge they already have.

Networking Skills

Once you have mastered your computer skills, you'll need to improve your skills with networking. It's important to know how a network functions and how to tweak it to make it better. The skills mentioned in this section are important to know; DNS, NAT, subnetting, DHCP, IPv4, IPv6, and routers and switches are all things you need to know about. You can learn many of the skills addressed in this section online.

As previously mentioned, oftentimes, amateurs are unaware of the different networking skills they will need to build upon. They may learn one or two of the skills mentioned and then fumble while hacking if they come across a different network. Therefore, any hacker who wants to improve should be aware of the various networking skills they need to have.

Linux Skills

Hackers often use Linux as their operating system. In fact, most tools developed for hackers are only designed for the Linux operating system. Linux can help the hacker achieve his end goal, unlike Windows. So, it's always a good idea to learn Linux. Any professional hacker should be adept at using Linux to hack into a system and identify its vulnerabilities.

Wireshark

Wireshark is a packet analyzer that is an open source tool. It's used by hackers to troubleshoot any network issues, analyze software and communications protocols and also to develop certain protocols for the system.

Expert hackers are versed in utilizing this analyzer to create protocols with ease for the system they are hacking into.

Virtualization

Virtualization is the art of making a virtual version of anything, like a server, storage device, operating system or networking resource. This helps the hacker test the attack that is going to take place before making it live. This also helps the hacker check if he or she has made any mistakes and revise the attack.

Professional hackers use this skill to enhance the effect of the hack they are about to perform. This gives them a perspective on the damage they can do to the software while protecting themselves. An amateur hacker usually does not learn how to cover his tracks. A perfect example for this is the boy from Mumbai who released an episode of Game of Thrones from season 7. Had he covered his tracks better, he would have been able to protect himself. This is why it's important to learn all about virtualization.

Security Concepts

It's vital that a hacker learns about different security concepts and understands the changes made to technology. A person who has a strong hold on security will be able to control different barriers set by the security administrators for the system they are hacking into.

Learning skills like Secure Sockets Layer (SSL), Public Key Infrastructure (PKI), firewalls, Intrusion Detection System (IDS) and other skills are important for hackers to learn. If you're an amateur, it is advised that you take courses like Security +.

Wireless Technology

This is a technology everybody is familiar with – information is sent using invisible waves as the medium. If you are trying to hack into a wireless device, you have to understand the functioning of that device. Therefore, it's vital that you learn the following encryption algorithms: WPA2, WPA WEP, WPS and the four-way handshake. It is also pertinent to understand the protocol connections, authentication and restrictions that surround wireless technology.

Scripting

This is a skill that every hacker must possess, especially the professionals. If a hacker were to use the scripts written by another hacker, he or she would be discredited for that. Security administrators are always vigilant about any hacking attempt and will identify a new tool, which will help them cope with that attack.

A professional hacker needs to build on this skill and ensure that he or she is good at scripting. Amateurs often depend on the scripts written by other hackers. They may or may not understand the script, which would land them in big trouble.

Database

A database helps a user store data in a structured manner on a computer that can be accessed in various ways. If a hacker wishes to hack into a system's database, he or she would need to be adept at different databases and also understand their functioning. Databases often use SQL to retrieve information whenever necessary. Therefore, it's important to learn these skills before you decide to hack into a database.

Professional hackers always need to know their way around a database to ensure that they make no mistakes and avoid getting caught.

Web Applications

Web applications are software through which you can access the Internet via your browser (Chrome, Firefox, etc.). Over the years, web applications have also become a prime target for hackers. It is extremely advisable that you some spend time understanding the functioning of web applications, as well as the databases that back those applications. This will help you make websites of your own either for phishing or for any other use.

The skills mentioned in this chapter are most important for hackers to develop. Professional hackers work to improve these skills right from the beginning and therefore are adept at hacking into any system easily. It's important for amateurs to build up these skills.

Ethical Hacking Process

Like every IT project, ethical hacking should always be planned. You have to determine the strategic and tactical issues in the process. Regardless of what the test is (whether it be a simple password-cracking test or a penetration test on an application on the Internet), you must plan the process.

Formulating the Plan

It is essential that you get approval before you begin the ethical hacking process. You have to ensure that what you are doing is known and visible to the system owners. The first step to working on the project is to obtain sponsorship. You can connect with an executive, manager, customer or even with yourself if you are your own boss. All you need is to have someone who can back you up and sign off on your plan. Otherwise, there's a possibility that someone may call everything off, stating that they never permitted you to test the devices.

If you're testing the systems in your office, you need a memo from your boss that gives you permission to perform them. If you are testing for a customer, you must ensure that you have a signed contract that states the customer's approval. It is pertinent that you obtain written approval so your effort and time don't go to waste. Also, this way, you will learn more about what you need to do to ensure that you stay out of trouble.

It's essential to have a detailed plan, because if you make one mistake, the systems can crash. However, this doesn't mean that you need to include the different testing procedures you intend on using. A well-defined plan or scope should include the following information:

•   Which systems need to be tested

•   The risks involved

•   When the tests will be performed and how long they will run for

•   How the tests will be performed

•   How much knowledge you have about the systems

- What you'll do if you come across a major vulnerability

- The deliverables, like security-assessment reports, high-level reporting of general vulnerabilities that the company should address and countermeasures that the organization should implement

You must always begin the testing with the most vulnerable and critical systems. For example, it is best to start with social engineering attacks or test computer passwords before you move on to more detailed issues. It's always a good idea to have a contingency plan in mind if something goes awry. There's a possibility that you may take the firewall down when you are assessing it, or you may close a web application while testing it; this will reduce employee productivity and system performance since the system would then be unavailable for use. There have been times when this has led to bad publicity, loss of data and loss of data integrity.

You should handle DoS and social engineering attacks carefully. You have to determine how these attacks will affect the system you're testing and the organization. You must also carefully determine when the tests should be performed. Do you want to test during business hours? Would it be better to test the systems early in the morning or late at night to avoid affecting the production of employees? Is it ideal to involve the people in the organization to be certain that they approve of the timing?

You have to remember that crackers do not attack your system during a limited period. Therefore, you should also use the unlimited attack approach. In this approach, you can run any type of test aside from social engineering, physical and DoS tests. You should never stop with one security hole since that will lead to a false sense of security. You have to continue to test to see what other vulnerabilities you can discover. This doesn't imply that you should continue to hack until all your systems crash. You should simply pursue the path you are on and hack until you can no longer hack the system.

One of the goals you should keep in mind when you perform these tests is to ensure that nobody detects the attack. For instance, you can perform your tests on a remote system or from a remote office when you're trying to avoid letting system users know what you're doing. If the users are aware of what you're up to, it will affect the outcome as they will then be on their best behavior.

You must be confident that you understand the system well enough to perform the hack. This will help to ensure that you protect the systems when you are testing them. If you are hacking your own system, it's not difficult to understand it. If you are hacking a customer's system, you will need to spend some time trying to understand how the system functions. Customers will never ask you to give them a blind assessment, because people are scared of these assessments. You should base all the tests you want to perform on the customer's needs and these assessments.

Selecting Tools

As with any project, you have to select the right tools if you want to complete the task successfully. That being said, you will not necessarily identify all the vulnerabilities in the system simply because you use the right tools. You must know the technical and personal limitations of your customer. Many security-assessment tools generate negative outcomes and false positives. Some tests won't locate the vulnerabilities. For example, if you perform a social engineering test or a physical-security test, it is easy to miss some weaknesses.

Certain tools focus only on specific tests, but no one tool can be used for everything. You cannot use a word processor to scan the network for any open ports, because that does not make sense. It is for this reason that you need specific tools for the test you wish to perform. Your ethical hacking efforts become easier when you have more tools at your disposal.

However, it is vital that you remember to choose the right tool for the task. You need to use tools like pwdump, LC4 or John the Ripper to crack passwords. SuperScan, which is a general port scanner, will not crack all passwords. If you want to perform an in-depth analysis of a web application, you should use tools like WebInspect or Whisker since they're more appropriate when compared to network analyzers like Ethereal.

When you need to select the right tools for a task, you should ask for advice from other ethical hackers, or you can post your questions on online forums and decide on the best tool to use.

Another option is using security portals like SearchSecurity.com, SecurityFocus.com and ITSecurity.com, or a simple Google search, to learn more about the different tools available for your tests. Experts provide their feedback and also give insights on the different types of tests an ethical hacker can perform.

Let's look at a list of some freeware, open-source and commercial security tools:

- Nmap
- EtherPeek
- SuperScan
- QualysGuard
- WebInspect
- LC4 (formerly called L0phtcrack)
- LANguard Network Security Scanner
- Network Stumbler
- ToneLoc

We will learn more about some of the tools listed above over the course of the book when we look at different types of hack attacks. Most people often misunderstand the capabilities of these hacking and security tools. This is because of the incorrect assumption that tools like Nmap (Network Mapper) and SATAN (Security Administrator Tool for Analyzing Networks) have gained bad publicity.

Some of these tools are complex, and you should familiarize yourself with each before you begin to use them. Here are some ways to do just that:

• Read the online help files or the readme files for the tools.

• Go through the user guide for any commercial tool.

• Join an online or formal class to learn more about the tool.

Executing the Plan

You need to be patient and have enough time on your hands to perform the hack. You also have to be careful while performing the hack. An employee looking over your shoulder or a hacker in the network will always watch what's going on, and this person will use the information they have obtained against you.

You cannot expect to perform an ethical hack when there are no crackers in the network, because that does not happen. You have to ensure that you keep everything private and quiet. This is critical when you are deriving, transmitting and storing the results of the test. You should try to encrypt these files and emails using tools like Pretty Good Privacy (PGP) and others. The least you can do is to protect the files using a password.

You're on a mission to get as much information as you can about the system you're testing. This is what a cracker will do. You should begin with a broad perspective and then narrow your focus:

1.   Look for the name of the organization, computer, network system and the IP Address; this information will often be available on Google.

2.   Now, narrow the scope and target the systems that you're testing. A casual assessment will turn up some information about the system, regardless of whether you are assessing web applications or physical-security.

3.   Narrow the focus with a critical eye and perform an actual scan. You should also perform detailed tests on the system.

4.   If you want to perform an attack, do it now.

Evaluating the Results

You should now assess the results of your hack to identify what you've discovered. It's advised that you make the assumption that these vulnerabilities were never uncovered before; this is where the results count. You need more experience to evaluate the results and identify the correlation between the vulnerabilities, and then you'll know your systems better than anybody else. This will make the evaluation process simpler going forward. The final step is to submit a formal report to your customer or to the upper management and outline your results. You must always keep both parties in the loop to show them that their money was well spent.

Moving On

When you have finished the ethical hacking test, you'll need to implement the analysis and also give the customer some recommendations. This will help to ensure the security of your systems. When you run these tests, new security vulnerabilities will appear. The information systems will always change, and these will become more complex. You'll uncover new hacker exploits and more security vulnerabilities as time goes on.

A security test is a snapshot of how secure your systems are. You should remember that things can change at any time, especially when you add a new system, apply patches or upgrade the software. That's why it's important to have a plan by which you perform regular tests to assess the system's security.

Phases of Ethical Hacking

As mentioned earlier, there is a set process that you should follow before you begin to hack a system or network ethically. This chapter covers the different phases of the ethical hacking process, which will help you or any other ethical hacker plan an attack. Every organization or company has a security manual that will explain the process differently. Most certified ethical hackers follow the six phases that will be discussed in this chapter.

Reconnaissance

The first phase of the process is the reconnaissance phase. This phase is also called the information-gathering phase. It's in this phase that the hacker should collect as much information as they can about the target system or network. Information is often accumulated about the following groups:

1.   Host

2.   Network

3.   People

Reconnaissance can be categorized into two types: active and passive.

Active Reconnaissance

In this type of reconnaissance, the hacker will interact directly with the target system or computer to gather information. For example, the hacker can use the Nmap tool to scan the network or system.

Passive Reconnaissance

In this type, the hacker will try to gather information about the system or network without interacting directly with the network, collecting data from websites, social media, etc.

Scanning

In this phase, the hacker will need to probe the target system and look for any vulnerabilities that it can exploit. For this purpose, the hacker can utilize Nexpose, Nessus and the Nmap tool.

Getting Access

Once the hacker identifies a vulnerability in the system or network, he or she will need to exploit that vulnerability to find out if they can enter the system. For this purpose, most hackers use a tool called Metasploit.

Maintaining Access

Once the hacker has gained access into the system, he or she will need to install a back door, or trap door. This will allow the hacker to enter the system whenever required in the future. Most prefer to use Metasploit during this phase.

Clearing Tracks

This is the unethical part of the process where the hacker will need to delete the log of every activity that they performed during the process of hacking the system.

Reporting

The last phase of the ethical hacking process is reporting. In this phase, the ethical hacker will need to prepare a report with all their findings and will also need to specify the different tools and methods that were used to perform the hack. The report should include the vulnerabilities found in the system and also list the solutions that the hacker wants to implement.

It is important to remember that the phases mentioned above are not set in stone. As an ethical hacker, you can always change the process or use different tools. You need to ensure that you are comfortable with the process. As long as you achieve the results you are looking for, you don't have to worry about sticking to the steps mentioned in this chapter.

Developing the Ethical Hacking Plan

As mentioned earlier, it's important for an ethical hacker to plan his or her efforts before they begin their task. You don't have to create a detailed plan but should provide information regarding what you're going to do as a part of the exercise. You should mention why it's important to perform the ethical hack and structure the process well.

Regardless of whether you are testing a group of computers or a web application, you must mention your goal and define the scope of your test. You should also determine the standards you'll be using to test the product. When you have written the plan down, you should gather different tools and familiarize yourself with those tools. This chapter will provide information on how you can create an environment that will improve the ethical hacking process to ensure that you're successful.

Getting the Plan Approved

It's important to get the plan approved, and the first step to do so is obtaining sponsorship. This approval should come from an executive, a customer, a manager or yourself. The testing may be canceled otherwise, or someone may deny authorization for these tests. There are times when there can be legal consequences for any unauthorized hacking. You have to ensure that you know what you're doing and that all your actions are visible.

This permission can be a simple memo from the senior management if you're running these tests on your systems. If you perform these tests for a customer, you should have a signed agreement in place with the customer's permission and authorization. It's important to obtain written approval to ensure your time and effort don't go to waste.

If you have a team of ethical hackers or are an independent consultant, you should purchase professional liability insurance from agents who specialize in business insurance coverage. This type of insurance is expensive, but it's very important to have the coverage.

Determining What Systems to Hack

You probably wouldn't want to evaluate the safety of all your systems at once. It could lead to more problems and is a difficult task. This isn't to say that one shouldn't eventually check every computer and application that's present; rather, it's suggested that when the time comes, one should break down their ethical hacking tasks into smaller tasks to ensure that it is easy to manage.

You may decide the systems you want to test depending on the risk analysis and answers to questions like:

•   Which are your most important systems?

•   If a system is hacked, what would be the biggest loss or lead to the most trouble?

- Which system is most vulnerable to attacks?

- Which systems are not strongly administered?

After the goals have been established, you can decide what systems need to be tested. This step helps one to carefully plan out their ethical hacking so that each person's expectation is established and to be certain the time and resources required for the job can be estimated as well.

The list mentioned below includes applications and systems that you should consider executing the hacking tests on:

- Firewalls

- Routers

- Network infrastructure as a whole

- Wireless access points and bridges

- Applications, Web servers and database servers

- Workstations, laptops, and tablet PCs

- E-mail and file/print servers

• Mobile devices (such as PDAs and cell phones) that have confidential information

• Client and server operating systems

• Client and server applications, such as e-mail or other in-house systems

Selecting the systems to test depends on several factors. If the network you're working on is small, everything can be tested from the get-go. It is better to test hosts that are open to the public such as web servers, emails and other associated apps. Hacking is flexible and all decisions should be made based on things that make the most sense business wise.

The first places to start are the the most vulnerable spots. You should consider the following questions:

• Where on your network does your computer or application reside?

• Which apps and operating system does it run?

• What type of important information is saved on it?

If the system that's being hacked is your own or a customer's, a previous security-risk assessment or vulnerability test would have generated this information. If this has been done, such documentation will help to point to systems needing further testing.

Ethical hacking is always a few steps above the higher-level information risk assessments and vulnerability testing. You should first get information about all the systems, including the entire organization, and then assess the systems that appear to be the most vulnerable.

It's ideal to begin with systems that have the best visibility. It makes more sense for you to focus on a file server or database that stores customer or any other critical or information. You can then focus on web servers, applications or firewalls after that.

Timing

It's often said that it's all about the timing. This is particularly true for an ethical hacker. While these tests are being performed, disruptions to any information systems, businesses and people must be minimal. Certain situations should be avoided at all costs, such as using the wrong timing for tests. Triggering a DoS attack in the middle of the day against a full-fledged e-commerce site or compelling yourself or others to perform tests to crack passwords at ungodly hours is a bad idea. Believe it or not, a 12-hour time difference can make a lot of difference! Every person involved must accept the complete timeline before you start. This helps everyone start together and thus set the right expectations.

Internet Service Providers (ISP) or Application Service Providers (ASPs) that are involved must be notified before any tests are performed on the Internet. This way, ISPs and ASPs will be aware of the tests that are taking place, and thus will minimize the chances that they will block your traffic if a malicious behavior is suspected and starts showing up on their firewalls or Intrusion Detection Systems (IDSs).

Conclusion

We live in a maddening world. Things seem to be going faster and faster, with apparently no option to keep anything under our control. We're forced to abandon old ideas, principles and habits as we're engulfed by cyberspace. More and more, we're living online, where large companies reign and have no intention of relinquishing the throne. In fact, they're increasingly seizing what used to be public domain and this goes unnoticed, without mainstream media fanfare or even a constructive mention in the news.

Smart machines and digital brains decide on whether our behavior is kosher and punish us within an instant. Any appeal is a formality and protests are memory-holed. Weak and powerless, you can keep shouting at the void, filled with resentment or you can learn to adapt. The sad truth is that there's simply no defeating the tech giants, so we have to wait for them to die a natural death.

Big companies collude with one another to keep a tight grip on the market and emerging technologies. As small startups rise to fight them, they too get absorbed and have their potential added to the conglomerate. Individuals have no say in the matter, as CEOs are insular, detached from reality and stubbornly dedicated to increasing company revenue. No amount of protesting, yelling or punishment will make them pay attention to the needs of the little guy. But there is a ray of light in all of it, a glimmer of hope as that same stubbornness makes them blind to small, consistent ethical hacking efforts by individuals.